Postharvest Diseases

Postharvest Diseases
Development and Cure

Sam Noble
Aswathy R V

2026
Agri Horti Press
New Delhi-110 084

ISBN: 978-93-49690-96-7 (Hardbound)

Published By

Agri Horti Press

New Delhi

E-mail: agrihortipress@gmail.com

Digitally Printed at : Replika Press Pvt. Ltd.

Preface

Postharvest Diseases: Development and Cure offers a detailed exploration of the complexities surrounding the diseases that affect crops after harvest. As the global demand for fresh produce continues to rise, the management of postharvest diseases has become an increasingly important issue in ensuring food security, minimizing waste, and maximizing the economic potential of agricultural products. This book delves into the scientific principles behind postharvest diseases, their development, and the innovative strategies for their prevention and cure.

Intended for students, researchers, agricultural professionals, and industry stakeholders, this book serves as both an educational resource and a practical guide. It presents a thorough understanding of the biology, epidemiology, and pathogenesis of postharvest diseases, offering insights into how environmental, handling, and storage conditions contribute to disease development in fruits, vegetables, grains, and other crops.

The early chapters of this book provide foundational knowledge on the types of pathogens—bacteria, fungi, viruses, and molds—that cause postharvest diseases. The book explains how these pathogens attack crops and the factors that make certain fruits and vegetables more susceptible to infections during the postharvest phase. It also covers the biochemical and physiological responses of crops to these diseases, helping readers understand how postharvest pathogens interact with the plant tissue and spread throughout the product.

Subsequent chapters focus on the mechanisms behind disease development, including the role of environmental factors such as temperature, humidity, and light in promoting pathogen growth. Postharvest diseases are often exacerbated by improper handling during harvesting, transportation, and storage. The book discusses best practices for handling crops, such as temperature control, humidity regulation, and sanitization, to reduce the risk of disease. It also explores how cultural practices and preharvest conditions influence the vulnerability of crops to postharvest pathogens.

A key theme in this book is the integration of innovative, sustainable methods for disease prevention and control. While traditional chemical treatments have been used to manage postharvest diseases, growing concerns about food safety, consumer preferences for organic products, and environmental sustainability have led to the development of alternative approaches. The book highlights non-chemical treatments such as biological control, the use of natural antimicrobials, and postharvest treatments like modified atmosphere packaging, ozone treatments, and hot water dips. Additionally, it explores the potential of novel technologies such as edible coatings, controlled-release packaging materials, and nanotechnology in extending shelf life and reducing disease incidence.

The book also emphasizes the importance of diagnostics and monitoring in postharvest disease management. Advanced techniques, including molecular diagnostics and sensor-based technologies, are discussed as tools to detect pathogens early and assess the quality of stored produce. These technologies enable more efficient and targeted interventions, helping to minimize postharvest losses and reduce the need for broad-spectrum treatments.

A significant portion of the book is dedicated to case studies and real-world applications. Examples from different parts of the world demonstrate how various crops—such as apples, tomatoes, bananas, and citrus—are affected by postharvest diseases, and the strategies that have been successfully implemented to mitigate these challenges. These case studies illustrate the interplay between scientific research, agricultural practice, and commercial interests in addressing postharvest disease problems.

Finally, the book addresses the broader implications of postharvest diseases for global food security. With a large proportion of harvested crops lost to diseases each year, the efficient management of postharvest diseases is critical for ensuring that food production systems can meet the growing demands of a global population. The book concludes by reflecting on the future of postharvest disease management, considering the role of research, policy, and international cooperation in tackling this persistent issue.

Postharvest Diseases: Development and Cure is designed to provide readers with a comprehensive understanding of the factors influencing postharvest diseases and the strategies available to combat them. By combining scientific insight with practical solutions, it aims to contribute to the development of more efficient, sustainable, and cost-effective methods for managing postharvest diseases, ultimately helping to reduce food waste, improve food security, and support sustainable agricultural practices.

Sam Noble

Aswathy R V

Contents

Chapter-1

Dissipation of Pre-Harvest Pesticides on 'Clementine' Mandarins after Open Field Application, and Their Persistence When Stored under Conventional Postharvest Conditions

1. Introduction

Citrus fruit are among the most widely produced and popular fruit in the world [1,2], due to their well-known health-promoting properties, their unique flavor, and versatility for combining with other ingredients to develop new foods. Citrus trees are very productive, and a normal 10 year old mandarin tree can produce 150–180 kg of fresh fruit. This productive capacity of citrus sustains the citrus fruit international business. The fresh fruit has a long half-life, which can be extended with cold storage, and the fruit can be sold overseas. Citrus are Uruguay's main fresh fruit export with sales in the international market as high as 110,000 tons per year. Orange, mandarin, and lemon represent 52, 34, and 14%, respectively, of the total fresh fruit volume exported, with Clementine the main mandarin cultivar exported [3]. Nevertheless, citrus plants are vulnerable to insects and fungal attack during the pre- and postharvest steps. In the field during the cropping season, citrus fruit are susceptible to insects such as *Phyllocnistis citrella* (citrus leafminer), the fruit fly, or fungi such as *Alternaria* spp. and *Phyllosticta citricarpa* (citrus black spot), which cause serious diseases that jeopardize overall productivity [4–6]. However, after harvest, the main citrus pest is green mold, *Penicillium digitatum*, which grows on the fruit even during cold storage [7,8].

To enhance productivity and protect the crop against plagues and pests, different pesticides are currently applied at different stages of the production chain. In the field, when the fruit is still developing, fungicides and insecticides from different chemical families are applied. Among the fungicides, the most widely employed are the phthalimides captan and folpet, benzimidazoles such as carbendazim, and tebuconazole, an inhibitor of ergosterol biosynthesis. The insecticides currently applied are chlorpyrifos, an organophosphate, and neonicotinoids such as imidacloprid or the chitin biosynthesis inhibitor buprofezin [9]. Pesticide residues and their persistence on or in food are an important and well-known concern for human health and environmental safety. When a pesticide is applied, it not only reaches the target but also reaches other organisms in the ecosystem. Pesticides dissipate within the environment after application. The processes of pesticide dissipation are complex and depend on their physicochemical properties, but also in their degradability by biotic and abiotic process [10,11]. After the banning of the most persistent organochlorines, efforts have been made to ensure the degradability and minimal bioaccumulation of newer pesticides. Once their protective action is accomplished, they should dissipate. As there could be more than one mechanism acting on pesticide dissipation, it is generally assumed that the kinetics of these processes follows a pseudo 1st-order equation that depends solely on the concentration of the compound under study. This assumption allows the determination of the half-life (time needed to reduce the concentration of the pesticide in the matrix under study to 50%), an important parameter that is used to establish the pre-harvest interval as well as the adjustment to accomplish the legal MRLs to ensure safe consumption [12]. A key issue for establishing accurate and safe pre-harvest intervals is to realize that pesticide dissipation depends not only on the chemical properties of the compound but also on the environmental conditions that rules their behavior. Temperature, moisture, rain, sunlight, and biota are the main parameters responsible for pesticide dissipation and they change from one region to another [11]. Many studies have reported the presence of imidacloprid [13,14], pyraclostrobin [15], and difenoconazole in oranges [16]. It is clearly established that the dissipation behavior of these compounds under field conditions depends on diverse factors, including plant species, chemical formulation, application method, weather conditions, and crop growth characteristics [17–19].

An extensive database of pesticide dissipation is available on the web. Lewis et al. [20] developed a dataset of pesticide dissipation rates in/on various plant matrices for the Pesticide Properties Database (PPDB). The influence of the specific part of the plant is highlighted, the leaves being the part that showed the greatest variations in the pesticide dissipation rates. A pesticide's mode of action plays a crucial role. Systemic pesticides suffer biotransformation and metabolism within the plant, but non-systemic ones can only be degraded by the epiphytic flora and the environmental conditions, which can change from one region to another and even from one cropping season to the next. The amount of epicuticular wax on the surface of the specific plant under study is also critical, and the amount depends on the water available to the plant. Water deficits trigger epicuticular wax biosynthesis, whereas epicuticular wax is thinner when the plant has good water supply. In addition, half-lives determined in greenhouses are different from those found in open field experiments. When the fruit are in greenhouses radiation is diminished, rainfall is nonexistent, while temperature can be controlled and, therefore, the pesticide half-life can change enormously [17,18]. Open field experiments are far more complicated to perform as the climatic conditions vary without any control, a situation that is avoidable in experiments performed in greenhouses. Another source of variation of pesticide half-life is postharvest treatment. Produce washing, hypochlorite baths or cold storage may have specific influences on the persistence of pesticides within the commodity. It has been demonstrated that washing reduced the lipophilic pesticide amount in leafy vegetables, but the same treatment had little effect in fruit with relatively high amounts of wax or hairs on their surface, where the pesticides dissolve and the lipophilic pesticide amount remains unchanged for longer periods. In general, cold storage diminishes pesticide dissipation by minimizing the viable microbiological flora, the lack of radiation inside the chamber and for slowing down all the chemical reactions that cause pesticide degradation. Consequently, dissipation studies for a given fruit in the specific growing conditions of the cropping region are necessary to determine if the established pre-harvest time ensures that residue levels are below the maximum residue limit (MRL). Several authors have reported on pesticide residues and their dissipation in citrus fruit, including pesticides such as flusilazole [21], 2.4 D [22], spirotetramat [23], spirodiclofen-pyridaben [24], and insect growth regulators [25]. Nevertheless, little attention has been paid to the effect of postharvest treatments on the level of the remaining residues of the pesticides applied in the field. In the case of the citrus production chain, fruit are washed and sanitized, followed by postharvest fungicide and protecting wax applications. The purpose of waxing is to impart a shiny appearance to the fruit and to reduce weight loss by slowing down senescence during storage [26]. The fruit thus treated is stored at 4 °C for some weeks and often shipped overseas. As has been pointed out, the artificial wax is another factor that influences the dissipation of pesticides from the fruit.

Among the crop-protecting agents employed in citrus production, pyraclostrobin and difenoconazole, two broad spectrum fungicides for citrus *Alternaria* brown spot control and citrus black spot control [27,28] have been intensively applied. Difenoconazole belongs to the largest fungicide family: the triazoles. Difenoconazole acts by inhibition of the demethylation step at C-14 of lanosterol during ergosterol synthesis [29]. By contrast, pyraclostrobin is a strobilurin fungicide. Its mechanism of action is based on binding at the oxidation site of quinol (Qo) (or ubiquinol site) of cytochrome b in the mitochondria and, therefore, it stops the transfer of electrons causing the inhibition of cellular respiration [30]. These fungicides also are transported within the plant. Difenoconazole is a systemic fungicide that is absorbed and translocated in a plant, whereas the strobilurins are quasi-systemic, as they diffuse translaminarly within a leaf, but do not spread throughout an entire whole plant. The number of applications per season is 1 and 1–2 for difenoconazole and pyraclostrobin, respectively.

Both are applied at a rate of 1600 L ha^{-1} (20 mL 100 L^{-1}) with an atomizer to the crown. The insecticides imidacloprid and spinosad are commonly sprayed for citrus leafminer control [4]. Imidacloprid is a neonicotinoid insecticide that targets the nicotinic acetylcholine receptor (nAChR). The union of the insecticide to the receptor causes a hyper-excitation of the nervous system and the death of the insect [31]. This systemic insecticide is commonly applied by painting the trunk, by irrigation or by spraying with an atomizer to the crown. Imidacloprid is applied twice per season at a rate of 30–50 mL 100 L^{-1} (irrigation or spray) or 1–3 mL $tree^{-1}$ (painting the trunk). Spinosad is a biological insecticide derived from the fermentation of the Actinomycete bacterium *Saccharopolyspora spinosa* [32], and is a natural mixture of spinosyn A and spinosyn D [33]. The mode of action of spinosad is over the nicotinic receptor, but the inhibition of the transmission of the nervous impulse does not occur at the nicotinic receptor itself. It also targets the Gamma-Aminobutyric Acid (GABA) receptor at the ion channels. This efficient insecticide is allowed in organic production, giving organic farmers an important tool to protect their production within the rules. The number of spinosad applications per season is 1 or 2 at a rate of 200–300 L ha^{-1} (15 cc L^{-1}). The insecticide is applied using an atomizer to the crown.

Even though the residues and dissipation patterns of difenoconazole, imidacloprid, pyraclostrobin, and spinosad have been determined for apple, cabbage, cucumber, grapes and tomato, there are few studies regarding their dissipation in citrus species. The dissipation of imidacloprid in lemon and sweet orange was evaluated by Phartiyal et al. [34] and Singh et al., [35], but there are few records of the behavior of these four pesticides during mandarin pre-harvest field conditions.

The aims of this work were: (i) to evaluate the dissipation of difenoconazole, imidacloprid, pyraclostrobin and spinosad in Clementine fruit in field conditions; and (ii) to assess the influence of postharvest practices on the terminal residue levels of the four pesticides in whole fruit.

2. Materials and Methods

2.1 Reagents and Apparatus

Acetonitrile (MeCN) and ethyl acetate (EtOAc) of High-Performance Liquid Chromatography (HPLC) quality were purchased from Sigma-Aldrich (Steinheim, Germany) and Pharmco Products Inc. (Brookfield, CT, USA), respectively. A Thermo Scientific (Marietta, OH, USA) EASYpure RoDi Ultrapure water purification system was used to obtain deionized water. Formic acid (HCOOH) p.a. 88% was obtained from Merck (Darmstadt, Germany). Sodium chloride (NaCl) was obtained from Carlo Erba (Arese, Italy). Anhydrous magnesium sulfate ($MgSO_4$) was purchased from Scharlau (Barcelona, Spain).

High purity reference standards (>98%) were purchased from Dr. Ehrenstofer (Augsburg, Germany) and were stored at −40 °C. Stock standard solutions at 2000 μg mL^{-1} were prepared by dissolving the standards in acetonitrile or ethyl acetate and stored in glass vials at −40 °C. Working standard solutions were prepared by diluting the stock solutions with acetonitrile.

An ultrasonic bath Wisd WUC-A03H from Daihan Scientific Co. Ltd. (Seoul, Korea), a SL16 centrifuge from Thermo IEC HN-SII (Langenselbold, Germany), a vortex mixer Wisd VM 10 and a Turbovap® Biotage LV evaporator (Charlotte, NC, USA) were employed during the sample preparation step.

2.2 Field and Postharvest Experiment

The experiment was conducted with Clementine mandarin cv. 'Clemenules' (*Citrus clementina* Hort. ex Tan.), one of the most popular mandarins worldwide, a medium-sized, easy to peel mandarin, with a smooth and deep reddish orange rind color and parthenocarpic ability [36]. Pesticide applications were made to healthy adult trees of Clementine mandarin grafted onto *Citrus trifoliata* (L. Raf.) rootstock. The trial was conducted in Salto in the northern region of Uruguay at 31°16′45.47″S, 57°53′52.15″W and an elevation of 41 m. The experimental plots contained six trees per pesticide in a single row, with a tree size of 9 m^3 in rows oriented N-NE; S-SW. The tree spacing was 2 m in the row and 5.5 m between rows. To assess the dissipation pattern of the selected agrochemicals, commercial formulations were sprayed separately with a Stihl SR450 backpack sprayer until reaching the drip point (approx. 6 L $tree^{-1}$). Each pesticide was applied around the tree. The formulations were applied using the same Good Agricultural Practices (GAPs) procedures employed in citrus production and at the recommended label rates. Table 1 shows the pesticide, active ingredients and rates used. The trials were initiated 42 d before harvest.

Table 1. Pesticides evaluated and rate per 100 L. European Union (EU) and Codex Alimentarius Maximum Residue Levels (MRL) for each analyte in mandarin. Vol: volume.

Commercial Formulation/ Manufacturer	Active (g L^{-1})	Vol (cc)	EU MRL (mg kg^{-1})	Codex MRL (mg kg^{-1})
SCORE 250 EC/Syngenta	Difenoconazole 250	20	0.6	0.6
SPINGARD 35F/Compañía Cibeles SA	Imidacloprid 350	50	1	1
COMET/BASF	Pyraclostrobin 250	30	1	2
TRACER I/Rutilan SA	Spinosad 480	20	0.3 z	0.3 z

z Spinosad residue definition for plant and animal commodities: sum of spinosyn A and spinosyn D.

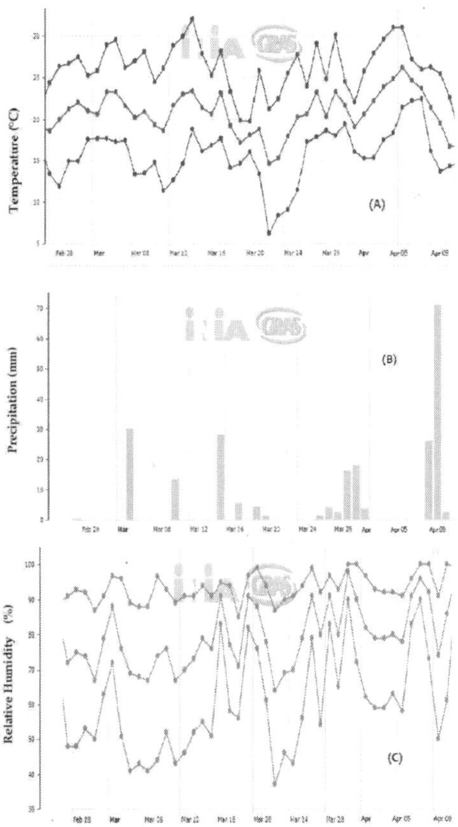

Figure 1. *Cont.*

Meteorological conditions from the spray date to fruit harvest were monitored. The daily temperature (24 h), precipitation (effective amount of precipitation), % relative humidity, and solar irradiation x relative Heliophany, under field conditions, were measured and are presented in Figure 1.

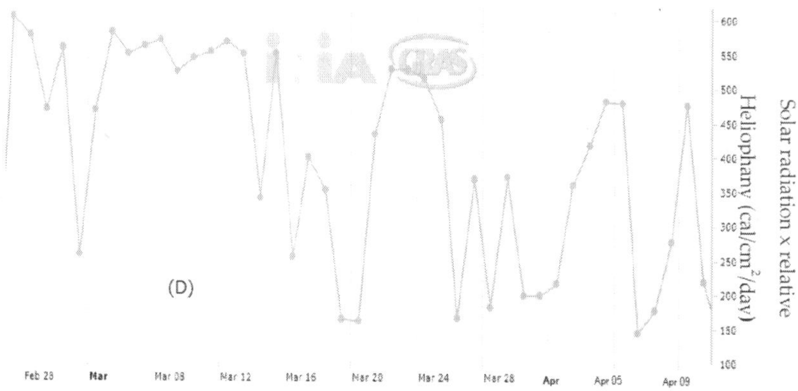

Figure 1. Maximum, average and minimum daily temperature (A), precipitation (B), relative humidity (C) and solar radiation x relative Heliofania (D).

After harvest, a postharvest treatment simulation was done in an experimental packing line at the National Agricultural Research Institute (INIA) (Salto, Uruguay). In the packing line, imazalil (FUNGAFLOR 75 PS, Janssen Pharmaceutica N.V.) at 1000 mg L^{-1} was applied in a cascade after hypochlorite sanitation and pre-drying steps. Then, a wax treatment (Brillaqua) was sprayed on the fruit. The treated fruit were air dried and stored for 32 day at 4 ± 1 °C and a relative humidity of 96±1% in an industrial cold chamber thus simulating the commercial conditions employed for citrus fruit that is exported (Figure 2). The relative humidity and temperature were controlled and recorded daily.

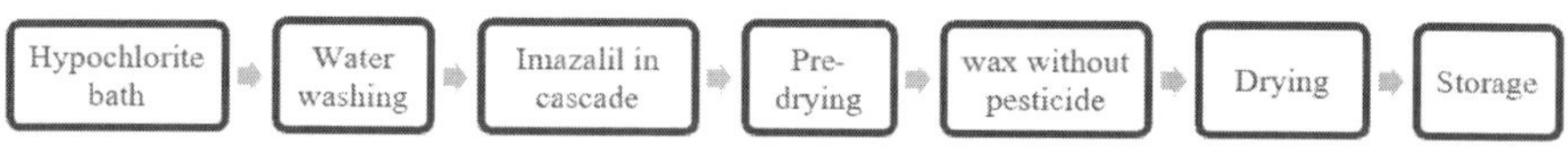

Figure 2. Packing-line pesticide application scheme (postharvest step).

2.3 Fruit Sampling

During sampling steps, a fruit blank was taken from untreated trees. From treated trees, sampling was performed following the Codex Alimentarius guidelines [37]. A total of 1 kg of fruit (approx. 9–10 fruit) was randomly taken by sampling from every part of the canopy (including the central part of the tree) from the six sprayed trees. The first sample was taken 1 h after pesticide application had dried and labelled as 0 day. Afterward, the fruit were collected randomly at 6, 21, 27 and 35 day after application of the pesticide and at normal harvest (42 day post spraying). Field samples were placed in polyethylene bags and transported on ice to the laboratory for analysis. Each fruit was cut into four pieces, with two opposite pieces selected for analysis. Each sample of two pieces was then homogenized and frozen (−18 °C) in individual polyethylene bags until analysis.

During the postharvest step, treated fruit were placed in plastic boxes and kept in the cold chambers. The individual fruit sampled were weighed before and after the cold storage, and an averaged weight of the fruit was recorded. For the analytical measurement, 1 kg of fruit (approx. 9–10 units) was taken randomly from the boxes. Samples were taken before and after the packing-line simulation and 32 day after cold storage; thereafter they were analyzed as described previously.

2.4 LC-MS/MS Operating Conditions

An Agilent 1200 liquid chromatograph (LC), with quaternary pump, degasser and thermostated autosampler coupled to a triple quadrupole API 4000 (4000 Q-TRAP ABSCIEX) was used for LC-MS/MS analysis. A Zorbax Eclipse XDB-C18 column, 150 mm long, 4.6 mm i.d. and 5 µm particle size, was kept at 40 °C during the chromatographic analyses. Water/ HCOOH 0.1% (*v*/*v*) (A) and acetonitrile (B) were employed as mobile phases at a flow of 0.6 mL min^{-1}. The operation of the LC gradient involved the following elution program: starting with 70% of component A, which was dropped to 0% in 12 min (3.5 min hold) then it rose back to 70% in 1.5 min (4 min hold) for a total run time of 21 min. 5 µL of each sample were injected.

MS/MS detection was performed in the multiple reaction monitoring (MRM) mode using an electrospray ionization (ESI) interface in the positive ion mode. The optimal MRM transitions, declustering potentials (DP) and collision energies (CE) of each studied compound were determined by infusing the standard solutions directly into the Q-TRAP with a syringe at a constant flow. A retention time window of 90 s was set for each analyte.

2.5 Residue Analysis and Method Validation

Pesticide residues were determined thrice for each sample, using an in-house validated method [38] and quantified by LC-MS/MS operating in MRM mode. Briefly, 10 g of homogenized sample was weighed into a 50 mL Teflon centrifugal tube. EtOAc (10 mL) was added, and the tube was manually shaken for 1 min. Afterwards, 8 g anhydrous $MgSO_4$ and 1.5 g NaCl were added. The tube was shaken vigorously by hand for 1 min, placed in an ultrasonic bath for 15 min, and centrifuged for 10 min at 3000 rpm. After phase separation, 2 mL of the organic phase was collected in a 4 mL flask. The extracts were concentrated to dryness under a N_2 flow in a 40 °C water bath. The residue was dissolved in 1 mL of MeCN [39].

Prior to analyzing the tissue samples, mandarin blanks (free of pesticide) were fortified at levels of 0.010 and 0.100 mg kg^{-1} by spiking a standard pesticide working solution in five replicate samples to determine recovery percentages. Precision was calculated as relative standard deviation (RSD) of the five replicates. An external matrix matching the standard method was used for quantification. The matrix effect (ME) was determined for each analyte as: (slope of calibration in matrix − slope of calibration in solvent)/slope of calibration in solvent. All figures of merit were determined.

2.6 Dissipation Models

The graphical representation of residue concentration decay over time is known as a pesticide dissipation curve. The residue data gathered from each sampling day and trial were averaged and a linear regression analysis of the data was performed following the procedure in [40]. This methodology allows the evaluation of different "model" equations which describe the dissipation of pesticide residues in agricultural products versus time.

For each pesticide, the experimental residue data was converted to achieve 1st-, 1.5th-, 2nd-order and 1st-order Root Factor (RF) model equations (Table 2). The transformed results were plotted versus time or versus root square time (RF model). The slope, intercept, and the coefficient of determination (R^2) of each linear regression were calculated.

Table 2. Linearizing transformations for the regression equation y = bx + a.

	Transformation	
Function	**For *y* (residue)**	**For *x* (time)**
1st-order	logR	none
1.5th-order	$1/\sqrt{R}$	none
2nd-order	1/R	none
1st-order RF z	logR	$\sqrt{t}$

z RF: "root function".

The selection of the best fit function from the four possible functions was achieved by comparing the deviations of the experimental values from the back-transformed curve. A modified coefficient of determination (r^2) and D quantity test were used for that purpose [41]. A test quantity D was calculated as:

$$D = /r/ - t\,(t^2 + (df))^{-0.5} \qquad (1)$$

where /r/ is the absolute value of the correlation coefficient (from transformed data) and t is the value of t, for n − 2 (degree of freedom: df), of the Student-t distribution table at the contrasted level of probability ($\alpha = 0.05$). If the calculated D quantity test was greater than zero, the correlation was confirmed. If the modified r^2 became negative, the fit was automatically regarded as not assured [40].

2.7 Data Management and Statistical Analysis

Data analysis was performed with Biosystems Analyst 1.5 software. Microsoft Excel Data Analysis tool was used for the statistical treatment. Student-test at the contrasted level of probability ($\alpha = 0.05$) was applied.

3. Results and Discussion

3.1 Method Performance

The main parameters including percentage of recovery, precision, linear calibration, limit of quantification (LOQ) and ME, were investigated according to the EU guidelines. The analytical method was developed to provide a rapid, accurate and efficient way to determine difenoconazole, imidacloprid, pyraclostrobin and spinosad residues in mandarins simultaneously. The results are listed in Table 3. The recovery percentages for the studied pesticides were within the Guidance document on analytical quality control and method validation procedures for pesticide residues and analysis in food and feed (SANTE) criteria (70–120%) [42].

Table 3. Percentage of recovery (Rec %), relative standard deviation (RSD %) at 10 and 100 μg kg^{-1} concentration levels in mandarin and matrix effects percentage (ME %).

			Concentration Level (μg kg^{-1})		
	10		100		
	Rec (%)	RSD (%)	Rec [1] (%)	RSD (%)	ME (%)
Difenoconazole	71.8 [z]	8.0	88.7	7.4	8.6
Imidacloprid	113.4	6.6	80.9	2.5	63.5
Pyraclostrobin	72.1	2.1	100.0	2.2	7.2
Spinosad	86.5	3.9	98.3	15.9	−55.1

z Average of n = 5.

3.2 Dissipation Studied under Field Conditions

Best Fitting Model

The best-fitted model equations for the four pesticides under field conditions on Clementine with their corresponding r^2 and D are presented in Table 4. The models which best predicts the experimental data forspinosad and difenoconazole were 1st-order equations. Imidacloprid and pyraclostrobin experimental data were best fitted by RF 1st-order and 2nd-order models, respectively. Dissipation curves from experimental data and back-transformed best-fitted models for difenoconazole, imidacloprid, pyraclostrobin and spinosad treatments are presented in Figure 3.

3.3 Initial Residues, Dissipation Dynamics and Postharvest Storage

The pesticide residue dissipation in the fruit in the field was studied and the depletion of the active principles was assessed. Similarly, the pesticide residues after the postharvest treatment were compared with those found at the end of cold storage. The fruit weight as well as the residue level of pyraclostrobin and difenoconazole remained unchanged after cold storage.

3.3.1 Difenoconazole

The initial residue level of difenoconazole was 0.060 ± 0.011 mg kg^{-1}, which was below the international settled MRL (see Table 1). Difenoconazole dissipated 80% 42 day after spray application. No significant differences were observed between the concentrations before and after the postharvest steps (Student t-test, $\alpha = 0.05$) (Figure 4).

Table 4. The 1st, 1.5th, 2nd, and Root Factor (RF) 1st order model equations that describe the dissipation of difenoconazole, imidacloprid, pyraclostrobin and spinosad in Clementine mandarin. The Best fitting model for each pesticide are in bold font.

	Difenoconazole	Imidacloprid	Pyraclostrobin	Spinosad
1st-order model				
Dissipation curve	$\mathbf{10^{-(0.016t+1.140)}}$	$10^{-(0.022t+0.334)}$	$10^{-(0.008t+0.591)}$	$\mathbf{10^{-(0.052t+0.744)}}$
r^2	0.857	0.883	0.890	0.987
D	0.198	0.197	0.124	0.003
1.5th-order model				
Dissipation curve	$1/(0.116t + 3.353)^2$	$1/(0.064t + 1.312)^2$	$1/(0.021t + 1.970)^2$	$1/(0.301t + 2.011)^2$
r^2	0.509	0.951	0.906	0.871
D	0.167	0.178	0.137	<0
2nd-order model				
Dissipation curve	1/(1.491t + 5.617)	1/(0.362t + 0.421)	1/(0.096t + 3.846)	1/(3.420t-1.062)
r^2	<0	<0	0.918	<0
D	0.13	0.124	0.149	<0
RF 1st-order model	√	√	√	√
Dissipation curve	$10^{-(0.103\sqrt{t} + 1.086)}$	$\mathbf{10^{-(0.148\sqrt{t} + 0.185)}}$	$10^{-(0.053\sqrt{t} + 0.535)}$	$10^{-(0.242\sqrt{t} + 0.658)}$
r^2	0.598	0.962	0.893	0.963
D	0.108	0.198	0.147	<0

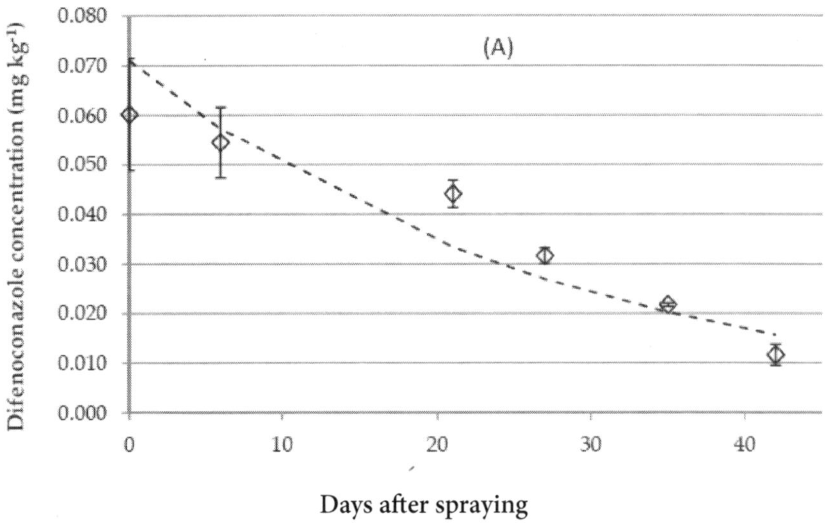

Figure 3. *Cont.*

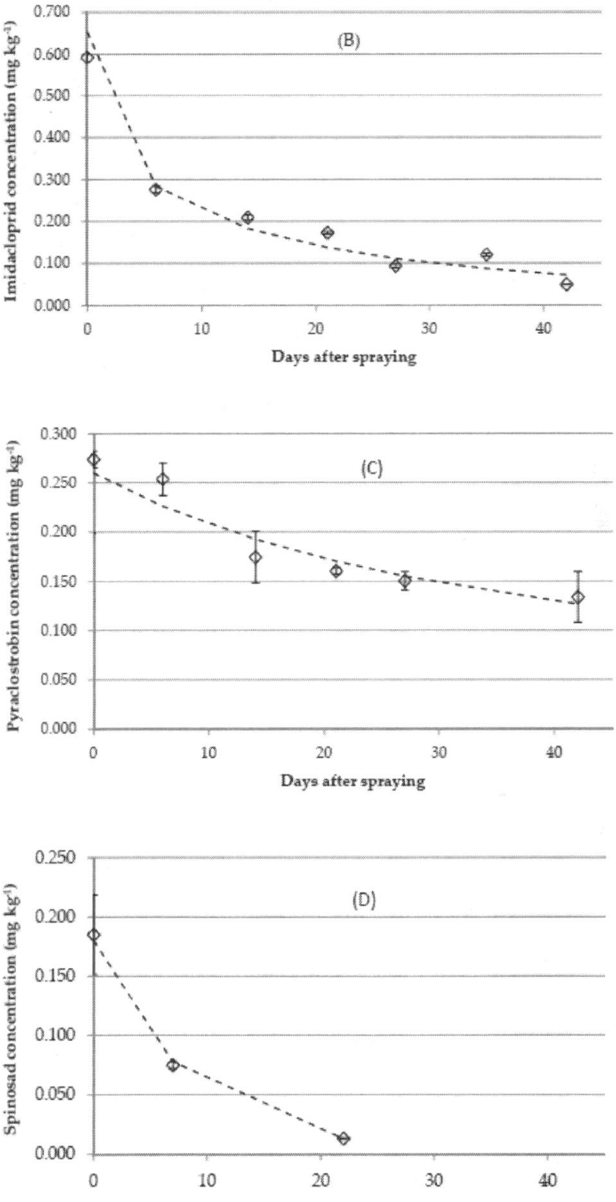

Days after spraying

Figure 3. Dissipation curves for difenoconazole (A); imidacloprid (B); pyraclostrobin (C) and spinosad (D) in Clementine mandarin in the days after spraying. -◊- Indicates experimental data. — Indicates the back-transformed dissipation curve (the best-fitted mathematical model: 1st-order equation for (A,D) Root factor (RF) 1st-order equation for (B) and 2nd-order for (C).

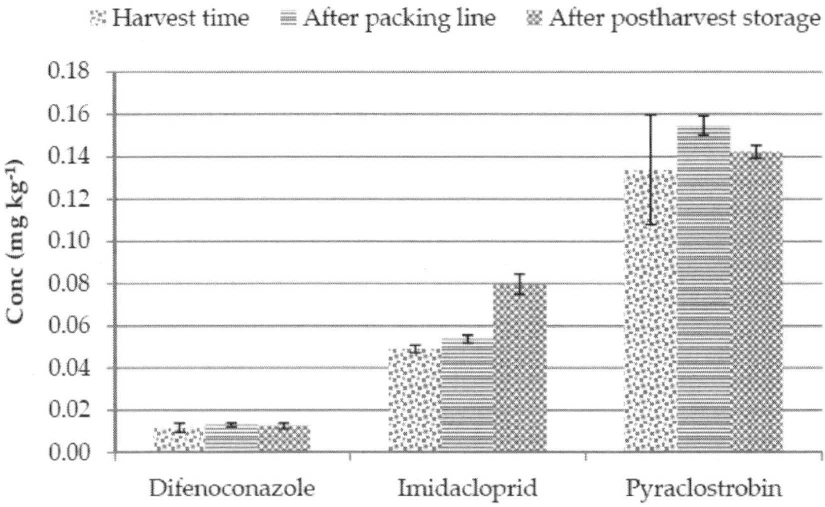

Figure 4. Difenoconazole, imidacloprid and pyraclostrobin residues on Clementine at harvest, after the postharvest packing line and after 32 d of cold storage. For difenoconazole and pyraclostrobin, no significant differences were detected between the pesticide levels. For imidacloprid residue concentration at the end of the postharvest period differed significantly from the residue level determined at harvest (Student t-test, α = 0.05).

3.3.2. Imidacloprid

The initial deposition of imidacloprid on Clementine was 0.592 ± 0.010 mg kg^{-1}. Imidacloprid residues dissipated 92% by harvest (42 day after spraying). The final imidacloprid residue concentration at the end of the postharvest period differed significantly from the residue level determined at harvest (Student t-test, α = 0.05). No clear conclusion can be made about the effect of cold storage on imidacloprid concentration. The higher levels of imidacloprid detected after cold storage were likely due to fruit water loss, a situation that was observed solely for the imidacloprid-treated mandarins and could be ascribed to some technical failure in the wax treatment. The dehydration of the fruit changed the level of co-extractives and, therefore, the matrix effect also changed. Changes in the matrix effect mean a different slope for the matrix-matched calibration curve, and therefore, in the residue quantitation. Our group established the presence of isobaric compounds that may have interfered with imidacloprid determination in citrus fruit that included mandarins [43]. If this interference was not compensated by the matrix-matched calibration, the imidacloprid levels would be overestimated, yielding higher concentration values for this compound.

3.3.3. Pyraclostrobin

At spraying time, the concentration of the studied strobilurin was 0.274 ± 0.009 mg kg^{-1}. A dissipation of 48% was observed for pyraclostrobin 42 day after spraying. This fungicide degraded more slowly than the other pesticides. The concentration after postharvest storage did not defer from that at harvest (Student t-test, $\alpha = 0.05$). Other studies demonstrated the storage stability of pyraclostrobin in commodities with high oil, water and acid content as well as in dry commodities for up to 18 months when stored deep frozen [44].

3.3.4. Spinosad

The residue of spinosad on Clementine at application time (day 0) was below the EU and Codex MRLs (0.3 mg kg^{-1}). Spinosad residues dissipated below the quantification limit (BQL) of 0.010 mg kg^{-1} at 28 day after application of treatment. At 22 day after spraying, 93% of this compound had dissipated.

The postharvest processes had little influence on the overall dissipation of the pesticides. No significant differences were detected between the pesticide level at the beginning and the end of the packing line. The level was almost unchanged after 32 day in cold storage, except for imidacloprid as explained above.

3.4 Half-Lives

According to the best fitting models, difenoconazole and spinosad dissipated following a pseudo 1st-order kinetics that only was a function of the concentration of the compounds. That means that the concentration declined exponentially and their half-lives, (the time needed to reach half of the initial concentration) remained constant and the clearance times, if needed, could be easily estimated. The half-life of difenoconazole and spinosad were 19.2 and 5.8 day, respectively. Imidacloprid and pyracostrobin half-lives determined with the best fitting models were 4.1 and 39.8 day, respectively. In these cases, the mathematical description of the pesticide dissipation were RF 1st-order and 2nd-order, respectively. These types of decay are concentration-dependent, meaning that the time needed to reach half of the concentration in the fruit (the pesticide half-life) will be a function of the pesticide level on the fruit at the starting time considered.

This is the first report of difenoconazole, spinosad and pyraclostrobin dissipation on Clementine mandarins. Imidacloprid showed almost the same half-life as that previously determined in the rind of sweet orange, which was 3.87 day [35].

Previous reports determined 3.5–5.3 day for the half-life of pyraclostrobin in fruit such as banana [45], strawberry [46], tomato [47], values that are 7 to 10 times lower than those reported in the present work (39.2 day). These results may partially be due to differences between the species, application doses, formulation, local environment, and crop growth characteristics [17–19]. It should be taken into account that citrus fruit have an important wax layer where due to the high octanol/water coefficient (Kow) of this pesticide, it could be retained. These facts could explain the deviation from a normal 1st-order kinetics behavior. It also must be taken into account that pyraclostrobin (pKow = 3.99) could dissolve in the essential oil vesicles of the citrus fruit. The dissolution of the pesticide in the oil sacs would hamper its direct contact with the environment, to which it dissipates. The pesticide that is not solubilized in the essential oil would dissipate quickly from the fruit surface and, afterwards, a slow release from the oil would change the speed of decay of pyraclostrobin.

The determined half-life for spinosad was 5.8 day. The short $t_{1/2}$ of this insecticide is influenced by light, especially the ultraviolet (UV) component [48]. Also, residue decline may be attributed to hydrolysis, biodegradation, or growth dilution [49]. Although there are no specific studies of dissipation of spinosad on mandarins reported in literature similar values of half-lives are reported for this pesticide in other vegetables and fruit. Values in the range of 3.5–3.9 day for zucchini [50], 1.7 day for tomato [51] and 1.4 day for cabbage [52] have been reported.

3.5 Estimating Storage Effect on Pesticide Level

The equations of the dissipation models allowed the estimation of the influence of the postharvest process through the comparison between the pesticide levels obtained after washing, imazalil and wax application and the final cold storage and the calculated ones after 32 d of postharvest storage. The expected concentration values, applying the best fitting models versus actual ones after cold storage at 74 d after spraying were: 0.005 versus 0.012 mg kg^{-1} for difenoconazole, 0.035 versus 0.080 mg kg^{-1} for imidacloprid and 0.091 versus 0.142 mg kg^{-1} for pyraclostrobin, respectively. Therefore, cold storage slowed down the dissipation of these pesticides.

4. Conclusions

The dissipation pattern of two fungicides and two insecticides on Clementine mandarin was studied. The dynamics of the decay of difenoconazole, imidacloprid, pyraclostrobin, and spinosad in field conditions was unique for each one. However, in all cases, a clear dissipation trend could be drawn. Spinosad decomposition was the most rapid, dissipating to below the quantification limit of 0.010 mg kg^{-1} 28 day after spraying. Pyraclostrobin showed the lowest dissipation rate under the field conditions. It was the highest persistence period reported for pyraclostrobin on fruit to date. For all four pesticides, the postharvest stage did not diminish the pesticide residue concentrations, which were below their MRLs before entering in the postharvest phase. These results support their safe use in Clementine cropping under the studied conditions.

References

1. Liu, Y.; Heying, E.; Tanumihardjo, S.A. History, Global Distribution, and Nutritional Importance of Citrus Fruits. *Compr. Rev. Food Sci. Food Saf.* **2012**, *11*, 530–545. [CrossRef]

2. FAO. FAOSTAT. Available online: http://www.fao.org/faostat/en/#data/QC/visualize (accessed on 6 September 2018).

3. DIEA. *Anuario Estadístico Agropecuario*, 20th ed.; MGAP: Montevideo, Uruguay, 2017.

4. Shinde, S.; Neharkar, P.; Dhurve, N.; Sawai, H.; Lavhe, N.; Masolkar, D. Evaluation of different insecticides against citrus leaf miner on Nagpur mandarin. *J. Entomol. Zool. Stud.* **2017**, *5*, 1889–1892.

5. Rodríguez, V.A.; Avanza, M.M.; Mazza, S.M.; Giménez, L.I. Pyraclostrobin effect to control of citrus black spot. *Summa Phytopathol.* **2010**, *36*, 334–337. [CrossRef]

6. Timmer, L.W.; Peever, T.; Solel, Z.; Akimitsu, K. Alternaria Diseases of Citrus—Novel Pathosystems. *Phytopathol. Mediter.* **2003**, *42*, 99–112.

7. Altieri, G.; Di Renzo, G.C.; Genovese, F.; Calandra, M.; Strano, M.C. A new method for the postharvest application of imazalil fungicide to citrus fruit. *Biosyst. Eng.* **2013**, *115*, 434–443. [CrossRef]

8. Lado, J.; Manzi, M.; Silva, G.; Luque, E.; Blanco, O.; Pérez, E. Effective alternatives for the postharvest control of imazalil resistant Penicillium digitatum strains. *Acta Hortic.* **2010**, *877*, 1449–1456. [CrossRef]

9. Boina, D.R.; Bloomquist, J.R. Chemical control of the Asian citrus psyllid and of huanglongbing disease in citrus. *Pest Manag. Sci.* **2015**, *71*, 808–823. [CrossRef] [PubMed]

10. Farha, W.; Abd El-Aty, A.M.; Rahman, M.M.; Shin, H.C.; Shim, J.H. An overview on common aspects influencing the dissipation pattern of pesticides: A review. *Environ. Monit. Assess.* **2016**, *188*. [CrossRef] [PubMed]

11. Navarro, S.; Oliva, J.; Navarro, G.; Barba, A. Dissipation of chlorpyrifos, fenarimol, mancozeb, metalaxyl, penconazole, and vinclozolin in grapes. *Am. J. Enol. Vitic.* **2001**, *52*, 35–40.

12. Karmakar, R.; Kulshrestha, G. Persistence, metabolism and safety evaluation of thiamethoxam in tomato crop. *Pest Manag. Sci.* **2009**, *65*, 931–937. [CrossRef] [PubMed]

13. Picó, Y.; El-Sheikh, M.A.; Alfarhan, A.H.; Barceló, D. Target vs non-target analysis to determine pesticide residues in fruits from Saudi Arabia and influence in potential risk associated with exposure. *Food Chem. Toxicol.* **2018**, *111*, 53–63. [CrossRef] [PubMed]

14. Chen, M.; Tao, L.; McLean, J.; Lu, C. Quantitative Analysis of Neonicotinoid Insecticide Residues in Foods: Implication for Dietary Exposures. *J. Agric. Food Chem.* **2014**, *62*, 6082–6090. [CrossRef] [PubMed]

15. Suárez-Jacobo, A.; Alcantar-Rosales, V.M.; Alonso-Segura, D.; Heras-Ramírez, M.; Elizarragaz-De La Rosa, D.; Lugo-Melchor, O.; Gaspar-Ramirez, O. Pesticide residues in orange fruit from citrus orchards in Nuevo Leon State, Mexico. *Food Addit. Contam. Part B* **2017**, *10*, 192–199. [CrossRef] [PubMed]

16. Ortelli, D.; Patrick, E.; Corvi, C. Pesticide residues survey in citrus fruits. *Food Addit. Contam.* **2005**, *22*, 423–428. [CrossRef] [PubMed]

17. Fantke, P.; Gillespie, B.W.; Juraske, R.; Jolliet, O. Estimating Half-Lives for Pesticide Dissipation from Plants. *Environ. Sci. Technol.* **2014**, *48*, 8588–8602. [CrossRef] [PubMed]

18. Fantke, P.; Juraske, R. Variability of pesticide dissipation half-lives in plants. *Environ. Sci. Technol.* **2013**, *47*, 3548–3562. [CrossRef] [PubMed]

19. Hanson, B.; Bond, C.; Buhl, K.; Stone, D. Pesticide Half-life Fact Sheet. Available online: http://npic.orst. edu/factsheets/half-life.html (accessed on 24 August 2018).

20. Lewis, K.; Tzilivakis, J. Development of a data set of pesticide dissipation rates in/on various plant matrices for the Pesticide Properties Database (PPDB). *Data* **2017**, *2*, 28. [CrossRef]

21. Wang, C.; Qiu, L.; Zhao, H.; Wang, K.; Zhang, H. Dissipation dynamic and residue distribution of flusilazole in mandarin. *Environ. Monit. Assess.* **2013**, *185*, 9169–9176. [CrossRef] [PubMed]

22. Chen, W.; Jiao, B.; Su, X.; Zhao, Q.; Sun, D. Dissipation and residue of 2,4-d in citrus under field condition. *Environ. Monit. Assess.* **2015**, *187*. [CrossRef]

23. Zhang, Q.; Chen, Y.; Wang, S.; Yu, Y.; Lu, P.; Hu, D.; Yang, Z. Dissipation, residues and risk assessment of spirotetramat and its four metabolites in citrus and soil under field conditions by LC-MS/MS. *Biomed. Chromatogr.* **2018**, *32*. [CrossRef] [PubMed]

24. Sun, D.; Zhu, Y.; Pang, J.; Zhou, Z.; Jiao, B. Residue level, persistence and safety of spirodiclofen-pyridaben mixture in citrus fruits. *Food Chem.* **2016**, *194*, 805–810. [CrossRef] [PubMed]

25. Payá, P.; Oliva, J.; Cámara, M.A.; Barba, A. Dissipation of insect growth regulators in fresh orange and orange juice. *Commun. Agric. Appl. Biol. Sci.* **2007**, *72*, 161–169. [PubMed]

26. Waks, J.; Schiffmann-Nadel, M.; Lomaniec, E.; Chalutz, E. Relation between fruit waxing and development of rots in citrus fruit during storage. *Plant Dis.* **1985**, *69*, 869–870. [CrossRef]

27. Vicent, A.; Armengol, J.; García-Jiménez, J. Rain fastness and persistence of fungicides for control of alternaria brown spot of citrus. *Plant Dis.* **2007**, *91*, 393–399. [CrossRef]

28. Miles, A.K.; Willingham, S.L.; Cooke, A.W. Field evaluation of strobilurins and a plant activator for the control of citrus black spot. *Aust. Plant Pathol.* **2004**, *33*, 371–378. [CrossRef]

29. Dong, F.; Li, J.; Chankvetadze, B.; Cheng, Y.; Xu, J.; Liu, X.; Li, Y.; Chen, X.; Bertucci, C.; Tedesco, D.; et al. Chiral Triazole Fungicide Difenoconazole: Absolute Stereochemistry, Stereoselective Bioactivity, Aquatic Toxicity, and Environmental Behavior in Vegetables and Soil. *Environ. Sci. Technol.* **2013**, *47*, 3386–3394. [CrossRef] [PubMed]

30. Balba, H. Review of strobilurin fungicide chemicals. *J. Environ. Sci. Health Part B* **2007**, *42*, 441–451. [CrossRef] [PubMed]

31. Simon-Delso, N.; Amaral-Rogers, V.; Belzunces, L.P.; Bonmatin, J.M.; Chagnon, M.; Downs, C.; Furlan, L.; Gibbons, D.W.; Giorio, C.; Girolami, V.; et al. Systemic insecticides (neonicotinoids and fipronil): Trends, uses, mode of action and metabolites. *Environ. Sci. Pollut. Res. Int.* **2015**, *22*, 5–34. [CrossRef] [PubMed]

32. Miles, M. The effects of spinosad, a naturally derived insect control agent to the honeybee. *Bull. Insectol.* **2003**, *56*, 119–124.

33. West, S.D.; Turner, L.G. Determination of spinosad and its metabolites in citrus crops and orange processed commodities by HPLC with UV detection. *J. Agric. Food Chem.* **2000**, *48*, 366–372. [CrossRef] [PubMed]

34. Phartiyal, T.; Srivastava, R.M. Dissipation of imidacloprid on lemon fruit rind under tarai agro-climatic condition of Uttarakhand. *J. Entomol. Res.* **2014**, *38*, 285–288.

35. Singh, N.; Bisht, S.; Yadav, G.; Kamari, B. Dissipation of imidacloprid on sweet orange fruits. *Int. J. Chem. Stud.* **2017**, *5*, 683–686.

36. Saunt, J. *Citrus Varieties of the World*; Sinclair Intl Business Resources: Norwich, UK, 2000.

37. Codex. *Recommended Methods of Sampling for the Determination of Pesticide Residues for Compliance with MRLs CAC/GL 33-199*; Codex: New Delhi, India, 1999.

38. Besil, N.; Pareja, L.; Colazzo, M.; Niell, S.; Rodríguez, A.; Cesio, V.; Heinzen, H. GC-MS method for the determination of 44 pesticides in mandarins and blueberries. In Proceedings of the 3rd Latin American Pesticide Residue Workshop, Food and Environment, Montevideo, Uruguay, 8–11 May 2011; p. 156.

39. Besil, N.; Pérez-Parada, A.; Cesio, V.; Varela, P.; Rivas, F.; Heinzen, H. Degradation of imazalil, orthophenylphenol and pyrimethanil in Clementine mandarins under conventional postharvest industrial conditions at 4 °C. *Food Chem.* **2016**, *194*, 1132–1137. [CrossRef] [PubMed]

40. Timme, G.; Frehse, H.; Laska, V. Statistical interpretation and graphic representation of the degradation behavior of pesticide residues II. *Pflanzenschutz-Nachrichten Bayer* **1986**, *39*, 187–203.

41. Timme, G.; Frehse, H. Statistical interpretation and graphic representation of the degradation behaviour of pesticide residues. 1. *Pflanzenschutz-Nachrichten Bayer* **1980**, *33*, 47–60.

42. European. *Guidance Document on Analytical Quality Control and Method Validation Procedures for Pesticides Residues Analysis in Food and Feed*; Document SANTE/11945/2015; European Commission: Brussels, Belgium, 2015.

43. Besil, N.; Cesio, V.; Heinzen, H.; Fernandez-Alba, A.R. Matrix Effects and Interferences of Different Citrus Fruit Coextractives in Pesticide Residue Analysis Using Ultrahigh-Performance Liquid Chromatography-High-Resolution Mass Spectrometry. *J. Agric. Food Chem.* **2017**, *65*, 4819–4829. [CrossRef] [PubMed]

44. EFSA. Modification of the existing MRLs for pyraclostrobin in various crops. *EFSA J.* **2011**, *9*, 2120. [CrossRef]

45. Yang, M.; Zhang, J.; Zhang, J.; Rashid, M.; Zhong, G.; Liu, J. The control effect of fungicide pyraclostrobin against freckle disease of banana and its residue dynamics under field conditions. *J. Environ. Sci. Health Part B* **2018**. [CrossRef] [PubMed]

46. Wang, Z.; Cang, T.; Qi, P.; Zhao, X.; Xu, H.; Wang, X.; Zhang, H.; Wang, X. Dissipation of four fungicides on greenhouse strawberries and an assessment of their risks. *Food Control* **2015**, *55*, 215–220. [CrossRef]

47. Jankowska, M.; Kaczynski, P.; Hrynko, I.; Lozowicka, B. Dissipation of six fungicides in greenhouse-grown tomatoes with processing and health risk. *Environ. Sci. Pollut. Res.* **2016**, *23*, 11885–11900. [CrossRef] [PubMed]

48. Adak, T.; Mukherjee, I. Investigating Role of Abiotic Factors on Spinosad Dissipation. *Bull. Environ. Contam. Toxicol.* **2016**, *96*, 125–129. [CrossRef] [PubMed]

49. Jacobsen, R.E.; Fantke, P.; Trapp, S. Analysing half-lives for pesticide dissipation in plants. *SAR QSAR Environ. Res.* **2015**, *26*, 325–342. [CrossRef] [PubMed]

50. Liu, Y.; Sun, H.; Wang, S. Dissipation and residue of spinosad in zucchini under field conditions. *Bull. Environ. Contam. Toxicol.* **2013**, *91*, 256–259. [CrossRef] [PubMed]

51. Ramadan, G.; Shawir, M.; El-Bakary, A.; Abdelgaleil, S. Dissipation of four insecticides in tomato fruit using high performance liquid chromatography and QuECHERS methodology. *Chilean J. Agric. Res.* **2016**, *76*, 129–133. [CrossRef]

52. Singh, S.; Battu, R.S. Dissipation kinetics of spinosad in cabbage (*Brassica oleracea* L.var. *capitata*). *Toxicol. Environ. Chem.* **2012**, *94*, 319–326. [CrossRef]

Chapter-2

Chitosan and Carnauba Wax Coatings are not Recommended for Yellow Carrots

1. Introduction

Postharvest losses of horticultural products restrict the availability of food in the world. With carrots, losses happen due to physical or esthetic flaws as a result of mechanical injuries, pathogen attack, shape defects, and withering [1,2]. Other studies led by our research group indicated that yellow carrots lose about 15% of fresh weight in only seven days when stored under ambient conditions (25 °C and 70% RH) (data not published).

Yellow carrots were recently embraced by Brazilian growers, appealing as a distinctive gourmet product, but one that does not have constant availability in the market. Better quality roots are obtained during a winter harvest and must be stored to maintain their availability for the longest time possible. However, this standard will only be achieved by adopting good postharvest practices, associated with the use of technologies that help to minimize losses and maintain the quality of the products for a longer period.

Edible coatings are considered the packaging of the future and may help reduce losses caused by withering (i.e., water loss. These coatings are thin layers of edible biopolymers made of proteins, polysaccharides, or lipids, which are applied directly onto a product surface, adhering to it as part of the final product. These materials create a barrier against physical injuries, microbiological contamination, loss or gain of moisture, and nutrient oxidation. Therefore, they help prevent product deterioration, extending storage life, sensorial quality, and safety. The difference between these coatings and plastic packaging is the fact that the former is edible and biodegradable, and may substitute partially or totally for the latter [3,4].

Chitosan is the basis of edible coatings from non-starch polysaccharides. It is made by alkaline deacetylation of chitin, which is a polysaccharide present in the exoskeleton of crustaceans. Besides the usual properties of coatings, the film generated by chitosan provides a good barrier to oxygen, although it is not an efficient barrier against carbon dioxide. It also contributes to the control of enzymatic browning and microbial activity, as well as the capacity of absorbing ions of heavy metals. These properties play a role in minimizing the oxidation process catalyzed by free metals and delay changes in the content of anthocyanins, flavonoids, and total phenolics [3,5,6].

Among lipid coatings, natural waxes, such as carnauba wax, stand out. They are chemically classified as esters of long chain aliphatic acids. As a non-polar compound, they are an efficient barrier to the exchange of water vapor. Carnauba wax has better emulsifying, lubricity, plasticising, and adhering properties than other waxes, besides a greater permeability to O_2. It also improves food appearance by providing gloss and improving superficial texture [7].

There are few studies that show the effects of edible coatings on carrot quality. Thus, our objective was to evaluate the use of different concentrations of two types of edible coatings, one of carnauba wax and the other of chitosan, on the conservation of the commercial quality of yellow carrots.

2. Materials and Methods

2.1 Raw Material and Treatments

'Yellow Stone' carrots were obtained from a commercial farmer (Araçoiaba da Serra, SP, Brazil, latitude, 23°30′19" S, and longitude, 47°36′51" O). Carrots were divided among seven treatments: (1) Control treatment (without coating application); (2) chitosan at a concentration of 1%; (3) chitosan at 3%; (4) chitosan at 5%; (5) carnauba wax at 0.5%; (6) carnauba wax at 1%; and (7) carnauba wax at 12%.

Chitosan (Nutrifarm do Brazil, São Paulo, Brazil, low molecular weight) was dissolved in a solution of 8% acetic acid and 1% glycerol. For applying coating, carrots were immersed in the solutions for 5 min and then kept on absorbent paper until the coating was fully dried. The carnauba wax with colophony (Aruá Comércio e Serviço LTDA., São Paulo, Brazil, BR12) was obtained at a concentration of 12% and diluted in distilled water for lower concentrations. Colophony is a wood rosin that acts as an emulsifier in carnauba wax. The wax application was made by using a cotton wool soaked in the solution. Then, carrots were laid on absorbent paper until the coating was fully dried. Finally, carrots were transferred to plastic containers (20 kg capacity) and stored at 2 °C, 95–100% RH for 30 days.

Analyses were performed on the day of coating applications (day 0) and after 7, 15, and 30 days. The study used a randomized experimental design with a factorial structure of 7 × 4 (treatments × days of analysis), with five repetitions of seven carrots (approximately 500 g) each.

2.2 Analisys

Brown stain index (BSI) was always performed by the same analyst by rating carrots based on visual appearance. Scores ranged from 1 to 3, being 1 = roots free of stains; 2 = roots with light or few stains; 3 = roots with a large amount and/or darker stains (Figure S1). The number of carrots that received each score was converted into a percentage and the result was expressed as % of roots. The shelf life limit was determined when there were no more carrots rated as 1. They were discarded at that time.

Color was evaluated by using a MINOLTA CHROMA METER CR-400 and the luminosity L*, a* and b* values were determined. Value a* is a red/green coordinate

(+a indicates red and −a indicates green) and value b* is a yellow/blue coordinate (+b indicates yellow and −b indicates blue) [8]. Three measurements were taken on the external surface of each of the seven carrots per repetition, one in the upper area, one in the middle area, and one in the lower area.

To determine the effects of these edible coatings on the carrots, periderm anatomical analyses were performed after 15 days of storage. An optical microscopy analysis was performed in both healthy and damaged carrots on cross-sectional cuts made with a sliding microtome Leica SN 2000 R (Leica Biosystems, Heidelberg, Germany), clarified in a bleach solution 10%, and washed with distilled water. Histochemical tests were conducted with Sudan IV for lipid substances [9] and double color with iodine green (greenish-blue color for lignin) and Congo red (red color for cellulose) [10]. Images were captured by a trinocular Leica DM LB microscope (Leica Microsystems, Wetzlar, Germany) connected to a Leica DC 300 F video camera. External images of the carrots were captured by aM205C (lens) microscope (Leica Microsystems, Wetzlar, Germany) connected to a Leica DFC 295 video camera.

The BSI and color results were submitted to a variance analysis (ANOVA) and the averages were compared by the Tukey test ($P \leq 0.05$), using the statistical software SISVAR, version 5.6 [11].

3. Results and Discussion

3.1 Visual Quality

To determine the effect of edible coatings on the quality of yellow carrots, changes in visual characteristics were observed. General appearance was taken as a marketing criteria. Overall, the use of edible coatings on yellow carrots was not viable, irrespective of treatment. The 12% carnauba wax 12% brown stains on the periderm right after the application and the intensity of these stains increased during storage (Figure 1). Carnauba wax at the lower concentrations (0.5% and 1%) and all chitosan coatings did

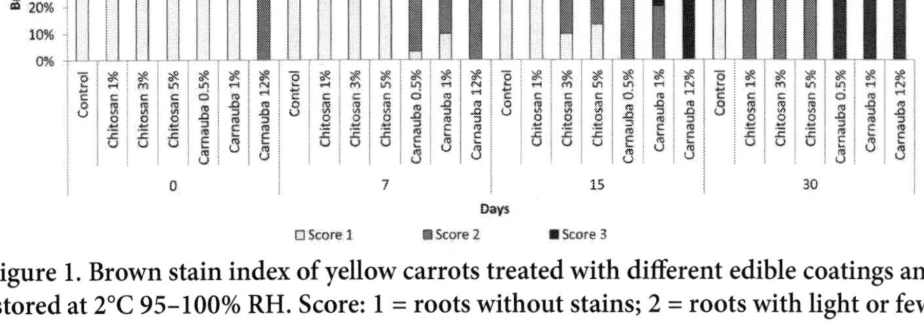

Figure 1. Brown stain index of yellow carrots treated with different edible coatings and stored at 2°C 95–100% RH. Score: 1 = roots without stains; 2 = roots with light or few stains; 3 = roots with a large amount of stains or with darker stains.

not cause stains on the day of application. Except for the control treatment, all treated carrots presented periderm stains on the seventh day of storage. On this day, all the carnauba treatments and 3% chitosan 3% presented more severe stains, receiving a score of 3. Treatments with 1% and 5% chitosan exhibited light or few stains after seven days of storage. On the 15th day of storage, carrots coated with carnauba wax were discarded due to the severity of the stains. For 1%, 3%, and 5% chitosan-coated carrots, only 27%, 10%, and 13% did not have stains, respectively. The control treatment exhibited 7% of the carrots with light stains. On the 30th storage day, despite the concentration, 100% of the carrots coated with chitosan were stained and thus not marketable.

The color of yellow carrots was affected by the application of coatings, confirming the data obtained from the visual rating. The luminosity decreased during cold storage for all treatments (Figure 2A). Except for 1% chitosan, all treatments presented lower luminosity values than the control samples. The decline in luminosity was proportional to the increase of the coating concentration. The reduction on the L* values indicated the darkening of the surface caused by the brown stains.

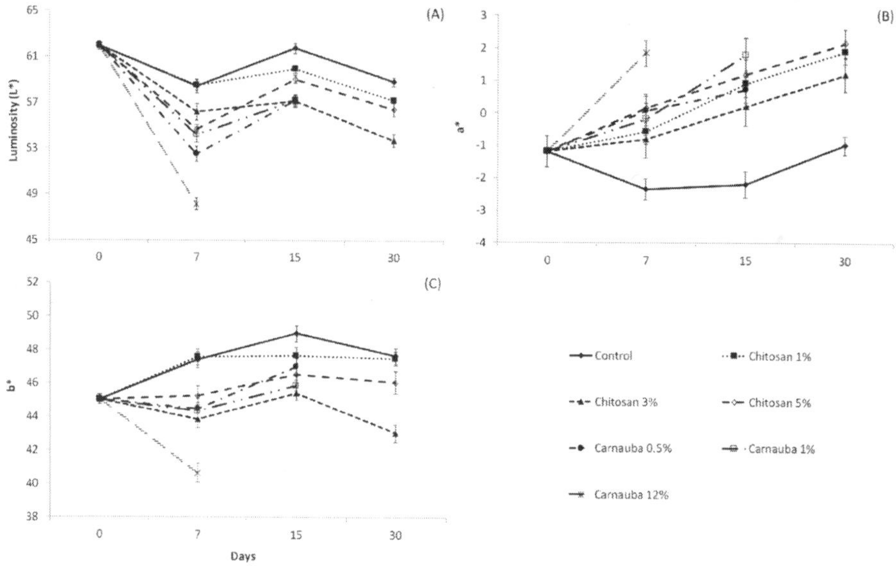

Figure 2. Luminosity (A) and values a* (B), and b* (C) of yellow carrots treated with different edible coatings and stored at 2°C 95–100% RH. Vertical bars represent the standard error of the mean (n = 5).

The control treatment did not show significant variation for the a* values during cold storage, always exhibiting the lowest values among the different treatments (Figure 2B). All of the other samples presented a* values that increased during storage, which indicated a change in the characteristic yellow color of this variety. The higher the a* value, the closer the sample is to a reddish hue. Brown stains in these treatments left a reddish tint to the carrots, which is directly related to the increase in the values of a*.

Control, 1% chitosan 1%, and 5% carnauba treatments showed an increase in the values of b* during storage. Chitosan at 1% interfered the least on the color of the yellow carrots, not differing from the control (Figure 2C). Carrots coated with 3% chitosan and 12% carnauba showed a reduction in the b* values from the beginning to the end of the storage period. This reduction showed a loss in the yellow color saturation, especially for the carrots coated with 12% carnauba, which also exhibited the highest degree of brown stains. Carrots coated with 5% chitosan and 1% carnauba 1% did not show any variation; however, their b* values were lower than the control.

3.2 Damage Evaluation with Microscope and Magnifier

Due to the observation of visual quality losses caused by the coatings, microscopy and magnifying glass techniques were used to investigate the extent of damage.

After 15 days of storage, damage affected the carrot periderm. Carnauba at 12% caused not only stains, but also depressions and superficial viscosity (Figure 3F). Carrots with chitosan presented lighter stains than the carnauba after 15 days of storage. However, at the end of the storage period, the stains were darker, similar to the ones caused by the carnauba.

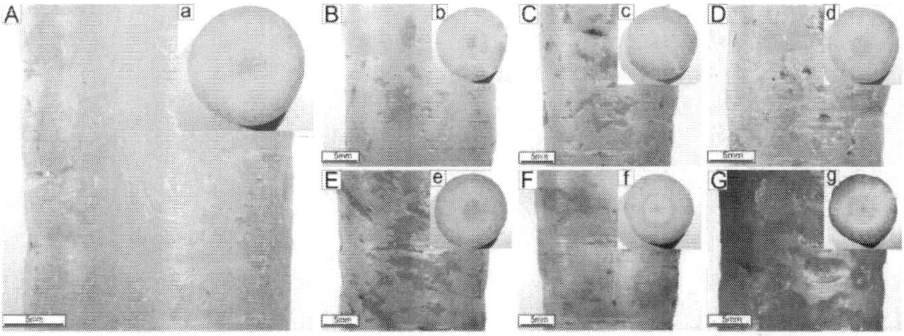

Figure 3. Details of yellow carrots 15 days after the application of different edible coatings and stored at 2°C 95–100% RH. (A,a) Control; (B,b) chitosan 1%; (C,c) chitosan 3%; (D,d) chitosan 5%; (E,e) carnauba wax 0.5%; (F,f) carnauba wax 1%; (G,g) carnauba wax 12%.

Carnauba wax at 6% (Figure 4A) was also tested (no data presented) and, like 12% carnauba (Figure 4B), it caused brown stains immediately after the coating application. The carrot surface has a periderm full of imperfections, creating a debarking aspect. Apparently, stains were concentrated on these imperfections along the carrot, suggesting that these structures were more sensitive to carnauba than the ones covered by the periderm. As the days advanced, stains spread and damaged the periderm cells, causing bigger injuries. Carnauba at 12% caused browning of the inner part on a significant area of the secondary phloem. The intensity of the injury also induced the occurrence of viscosity over the stains (Figures 3F and 4B) and depressions (Figure 4). Depressions were generated by the compaction of periderm and parenchyma cells (Figure 4C,E). In some cases, there was periderm rupture (Figure 4D,F) and deposition of lignin on the depressed region (Figure 4F). It was possible to observe that the damaged regions frequently occurred close to the secretory canals, which are located longitudinally in the root. This intensifies the depression and brings about the rupture and exposition of these canals (Figure 4F).

Brown stains probably resulted from the oxidation of phenolic compounds and the subsequent formation of darker compounds. The interaction between the coating and the cells seems to have degraded cell structures, and thus this decompartmentalization provided the contact between phenolic compounds and oxidative enzymes. The *o*-quinones, which are products of this reaction, have an undesirable brown color that stands out on the yellow carrot surface. It depreciates its appearance and commercial quality.

The emergence of stains on the treatments with carnauba wax may be related to the basic pH of the product (around 9.0) that remains unaltered after its dilution. Solutions that are very basic may cause injuries on the cell walls, such as burns. Caron et al. [12] also reported the emergence of stains on 'Brasília' carrots. These stains were caused by the application of carnauba coatings undiluted or in concentrations of 1:3 and 1:6, depending on the brand of coating used.

Based on these results, chitosan and carnauba wax coatings at the tested concentrations are not recommended for yellow carrots, as they damaged the carrots appearance, thus making them unmarketable.

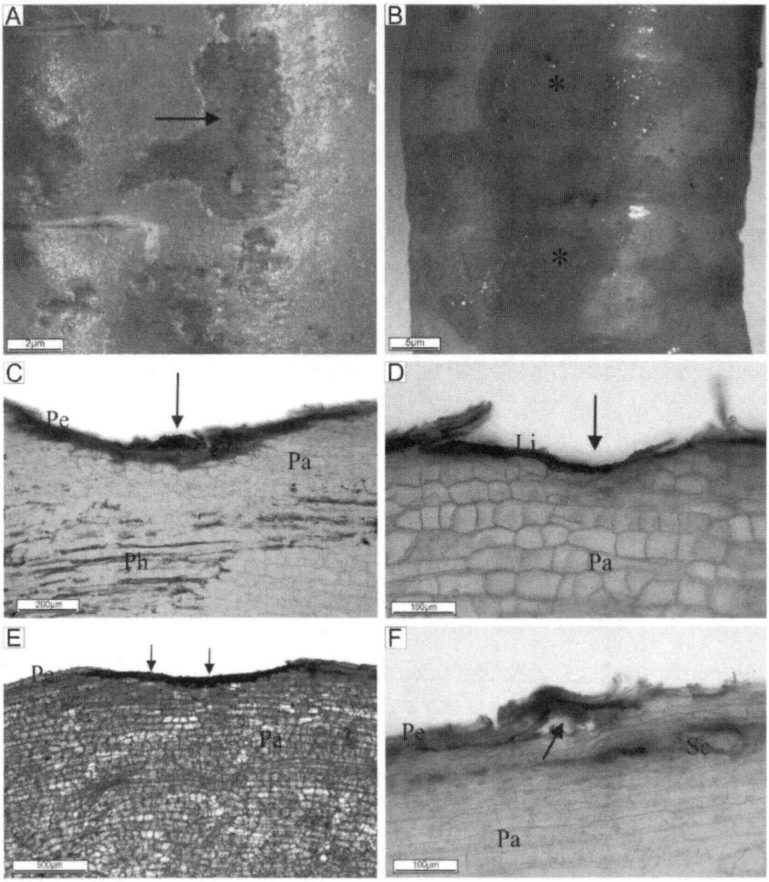

Figure 4. Damage to yellow carrots promoted by the carnauba wax: (A) Carnauba wax 6%; (B) carnauba wax 12%; (C) carnauba wax 6%; and (D) carnauba wax

12%. Photomicrographs of cross sections of superficial region of carrots: (C) Depression and compaction of the periderm. (D) Rupture and compaction of periderm cells evidenced by lignin deposition. (E) Compression of the periderm layers. (F) Periderm rupture near the secretory canal. Sc = Secretory canal; Ph = Phloem; Li = Lignin; Pa = Parenchyma; Pe = Periderm (arrows = depressions; * regions with viscous appearance).

References

1. Almeida, E.I.B.; Ribeiro, W.S.; da Costa, L.C.; Horácio, H.; de Lucena, J.A.B. Levantamento de perdas em hortaliças frescas na rede varejista de Areia (PB). *Revista Brasileira de Agropecuária Sustentável (RBAS)* **2012**, *2*, 52. [CrossRef]

2. Lana, M.M.; Moita, A.W.; Nascimento, E.F.d.; Souza, G.d.S.e.; Melo, M.F.d. Identificação das causas de perdas pós-colheita de cenoura no varejo, Brasília-DF. *Hortic. Bras.* **2002**, *20*, 241–245. [CrossRef]

3. Lacroix, M.; Le Tien, C. Edible films and coatings from nonstarch polysaccharides. In *Innovations in Food Packaging*; Han, J.H., Ed.; Academic Press: London, UK, 2005; pp. 338–361. [CrossRef]

4. Debeaufort, F.; Quezada-Gallo, J.-A.; Voilley, A. Edible Films and Coatings: Tomorrow's Packagings: A Review. *Crit. Rev. Food Sci. Nutr.* **1998**, *38*, 299–313. [CrossRef] [PubMed]

5. Nieto, M.B. Structure ad function os polyssacharide gum-based edible films and coatings. In *Edible Films and Coatings for Food Applications*; Embuscado, M.E., Huber, K.C., Eds.; Springer: New York, NY, USA, 2009; pp. 57–112. [CrossRef]

6. Dhall, R.K. Advances in Edible Coatings for Fresh Fruits and Vegetables: A Review. *Crit. Rev. Food Sci. Nutr.* **2013**, *53*, 435–450. [CrossRef] [PubMed]

7. Rhim, J.W.; Shellhammer, T.H. Lipid-based edible films and coatings. In *Innovations in Food Packaging*; Han, J.H., Ed.; Academic Press: London, UK, 2005; pp. 362–383. [CrossRef]

8. Minolta, C. *Precise Color Communication: Color Control from Feeling to Instrumentation*; Minolta Corporation Instrument Systems Division: Osaka, Japan, 1994; p. 49.

9. Jensen, W.A. *Botanical Histochemistry: Principles and Practice*; W.H. Freeman and Company: San Francisco, CA, USA, 1962; p. 408.

10. Dop, P.; Gautié, A. *Manuel de Technique Botanique*; Lamarre: Paris, France, 1928; p. 594.

11. Ferreira, D.F. Sisvar: A Guide for its Bootstrap procedures in multiple comparisons. *Ciência e Agrotecnologia* **2014**, *38*, 109–112. [CrossRef]

12. Caron, V.C.; Jacomino, A.P.; Kluge, R.A. Conservação de cenouras 'Brasília' tratadas com cera. *Hortic. Bras.* **2003**, *21*, 597–600. [CrossRef]

Chapter-3

Effect of Electronic Cold-Pasteurization™ (ECP™) on Fruit Quality and Postharvest Diseases during Blueberry Storage

1. Introduction

Blueberries (*Vaccinium* spp.) are becoming increasingly popular due to the rising awareness of the health benefits of consuming blueberry fruit, which include decreased risk of cardiovascular diseases, improved cognitive performance, and decrease in aging-related damage [1,2]. Commercially important blueberry species include lowbush (*Vaccinium angustifolium* Ait.) and northern highbush (*Vaccinium corymbosum* L.) mainly cultivated in the northern parts of the United States, and rabbiteye (*V. virgatum* Ait.) and southern highbush (hybrids of *V. corymbosum*, *V. virgatum*, and *V. darrowii* Camp.) grown mostly in the southern states [3,4]. Recently, production of blueberries has expanded to 27 countries (in 2011) compared with only ten countries in 1990 [5]. The United States is the largest producer of blueberries globally [5], supplying 347.7 million kg of cultivated and wild blueberries in 2016 [6]. The United States also plays an important role in the import and export trade of blueberries [7]. In 2016, the United States exported 31.7 million kg of fresh and 25.4 million kg of frozen blueberries and imported 149 million kg of fresh and 75.6 million kg of frozen fruit [8].

As global production and trade continues to rise, it becomes increasingly important to maintain fruit quality, nutrient content, phytosanitary safety, and eliminate pests and diseases in blueberries during storage to ensure that this fast-growing export and import market is not negatively impacted. Postharvest losses in fruits can vary from 10 to 40% [9]. After harvest, blueberries have a shelf-life of approximately 7 to 40 days depending on the genotype, method of harvest, and storage regime [9,10]. During postharvest storage, blueberry fruit quality can decline due to fruit softening [11]. Other contributing factors in loss of fruit quality are postharvest diseases caused primarily by fungal plant pathogens such as *Colletotrichum* spp. (ripe rot), *Alternaria* spp. (Alternaria fruit rot), and *Botrytis cinerea* (gray mold), among others [12–15]. In addition to postharvest disease-causing organisms, it is important to eliminate foodborne pathogens or associated indicator organisms [16–18]. Although outbreaks of foodborne illnesses associated with consumption of blueberry fruit have been relatively rare, produce brokers and buyers have begun to apply rigid (and typically proprietary) microbial standards to frozen blueberries destined for the processing market [19]. Although similar standards currently are not in place for the fresh-market, reducing microbial risk remains a key consideration for fresh-market production as well [20]. Finally, in order to export blueberries to other countries, they are required to be certified free of certain insect pests such as Mediterranean fruit fly (*Ceratitis capitata*), South American fruit fly (*Anastrepha fraterculus*), European grapevine moth (*Lobesia botrana*), blueberry maggot (*Rhagoletis mendax*), and plum curculio (*Conotrachelus nenuphar*) [21,22].

Fumigation of export goods with methyl bromide was the most commonly used phytosanitary treatment for elimination of pests, but has been phased out in the United States, with the exception of a few critical uses [23,24]. Methyl bromide also requires the produce temperature to be increased in order to be effective, thereby breaking the cold-chain and potentially having an adverse effect on quality. Interruption of cold-chain can decrease shelf-life considerably by increasing undesirable fruit metabolism [25]. Irradiation using gamma rays, X-rays, or electron beams could be an alternative to fumigation in eliminating pests and in preserving quality by reducing decay organisms and plant and human pathogens [23,24,26]. Previous work supported the use of electron beam and gamma irradiation to maintain shelf-life and fruit quality attributes in blueberry fruit [27–30]. In the United States, regulatory approval has been obtained for the use of irradiation on fresh fruits and vegetables up to 1 kGy [31]. Previous studies suggested an irradiation dose of 0.4 kGy to be effective against most insect pests, 0.2–0.8 kGy to cause a 1-log reduction in surface bacterial pathogens causing foodborne illness, and higher doses of 1–3 kGy for postharvest disease-causing fungi [22,32–34].

The objective of this study was to determine the effect of irradiating postharvest blueberry fruit using a new form of electron beam technology, Electronic Cold-Pasteurization™ (ECP™) developed by ScanTech Sciences (Norcross, GA, USA) at their Research and Development (R&D) facility at Idaho State University (ISU). This R&D facility is a small-scale version of a commercial ECP™ food treatment facility, which is currently being constructed by ScanTech in McAllen, TX and will be operational in the fourth quarter of 2018. This technology employs a highly focused beam of electrons, treating samples for only milliseconds on a high-speed conveyor while maintaining cold-chain integrity. A key advantage of electron beam irradiation over gamma rays (from nuclear sources such as Cobalt-60) or X-rays is the ability to deliver extremely high dose rates with improved accuracy since the beam dynamics can be more precisely controlled. These high dose rates equate to significantly less time for treatment and, consequently, potential for higher quality produce. The ECP™ treatment can treat an entire truckload (around 60,000 clamshells) of blueberries in a little over 30 min, whereas gamma rays can take several hours for the same quantity (C. Starns, unpublished observations). This is the first study to investigate the effect of irradiation on fruit quality attributes, postharvest disease incidence, and surface microbes of food safety concern in two southern highbush blueberry cultivars treated with ECP™ prior to cold storage.

2. Materials and Methods

2.1 Fruit Collection and Irradiation

Two trials were conducted with hand-harvested fruit from southern highbush blueberry cultivars 'Farthing' and 'Rebel' in Alma, GA. In trial 1 (April 2016), 'Farthing' fruit were obtained from a commercial packing facility, where fruit had already been prepacked into pint-size clamshell containers (473 mL). In trial 2 (May 2016), 'Rebel' fruit were obtained from a different packing facility, also already prepacked in pint-size clamshells. In addition, trial 2 included 'Farthing' fruit hand-harvested by the investigators from a commercial blueberry farm and packed into pint-size clamshells.

A subsample of clamshells in each trial was taken directly to the University of Georgia, Athens, GA, USA (330-km transit in refrigerated cooler) to serve as an unshipped control (not transported to and from the irradiation facility). Initial fruit quality attributes and postharvest disease incidence were recorded from this unshipped control. The remaining fruit in clamshells were arranged on standard flats (12 clamshells/flat), placed in a styrofoam cooler with ice packs, and shipped overnight from Alma, GA to ISU, Pocatello, ID. A foam sheet was placed on the inner side of the lid of each clamshell and in between clamshells to minimize fruit injury during shipment.

At ISU, fruit in clamshells were subjected to electron beam irradiation treatment at ScanTech's R&D facility using proprietary ECP™ technology. A 10-MeV electron beam, driven by an advanced high-energy electron accelerator, is magnetically focused through a scanning horn which delivers precision dose control. At the R&D facility, clamshells containing fruit were subjected to four levels of irradiation, 0, 0.15, 0.5, and 1.0 kGy; the treatments were completed in less than a second per clamshell. The respective doses were achieved using the National Institute of Standards and Technology (NIST)-traceable alanine pellets with extensive dose mapping on various blueberry configurations prior to the experimental fruit being shipped to the facility. Hundreds of data points were obtained and measured on a Bruker Bio-spin Electron Paramagnetic Resonance spectrometer, all of which are NIST traceable and International Organization for Standardization/American Section of the International Association for Testing Materials compliant. Treatments were replicated four times (i.e., four clamshells/irradiation level/postharvest storage period/cultivar), with a few exceptions where fewer replicate clamshells were available. The 0-kGy treatment served as an untreated control wherein fruit were shipped but not irradiated. After irradiation, fruit were shipped back by overnight courier to the University of Georgia where they were placed in a walk-in cooler at 2 to 4 °C under high relative humidity (>85%) until further assessment. The entire shipping and treatment process (from Alma to the treatment facility at ISU and to Athens for cold-storage and evaluation) took between 6 to 7 days. The unshipped control clamshells were stored in a 2 to 4 °C walk-in cooler until further evaluation. Fruit were removed from cold storage and evaluated for postharvest fruit quality attributes at 6, 13, and 25 days after irradiation treatment; microbial load on the fruit surface at 6 days after treatment; and postharvest disease incidence at 6 and 13 days after treatment followed by 4 days at room temperature. Fruit quality, microbial load and postharvest disease incidence analyses at a given time-point were performed using four replicates; for every replicate, fruit from a separate clamshell were used and divided for the above analyses.

2.2 Evaluation of Fruit Quality Attributes

For evaluation of fruit quality, visual assessment as well as measurement of fruit weight, texture, titratable acidity (TA), and total soluble solids (TSS) content were performed. For visual assessment, 30 fruit per replicate were scored for symptoms of bruising such as tears, dents, leakiness, or signs of mold. Fruit were examined by eye for visual defects and percent sound fruit were calculated. For fruit texture, two variables, fruit compression and skin puncture force, were measured on 12 fruit per replicate using a fruit texture analyzer (GS-15, Güss Manufacturing, Strand, South Africa); fruit were oriented on the equatorial plane for this assessment. For compression measurements, a 1.5-cm diameter plate was used with parameters set at forward speed 6 mm/s, measure speed 5 mm/s, and measure distance 1.00 mm. For skin puncture force measurements, a 1.5-mm flat-tip probe was used with parameters set at a forward speed 10 mm/s, measure speed 5 mm/s, and measure distance 3.00 mm. Fruit weight was recorded on 20 individual fruit per replicate using a balance (Quintix Precision Balance, Sartorius, Bohemia, NY, USA).

For TA and TSS measurement, juice was extracted from ~40 g of fruit per replicate using a household blender and centrifuged for 10 min at 3901X *g* using a benchtop centrifuge (Allegra X-22, Beckman Coulter Life Sciences, Indianapolis, IN, USA). The resulting supernatant was filtered through two layers of cheesecloth. To measure TSS, 300 μL of supernatant was tested using a digital handheld refractometer (Atago USA, Belleveue, WA, USA). For TA, the supernatant was titrated using an automatic mini titrator (Hanna Instruments, Woonsocket, RI, USA) and values were reported as percent citric acid (CA). Statistical analysis (one-way analysis of variance for a completely randomized design) was performed separately for each trial and cultivar using JMP Pro 12 (SAS Institute, Cary, NC, USA). Means were separated using Tukey's Honest Significant Difference (HSD) test ($\alpha = 0.05$).

2.3 Evaluation of Fruit Surface Contaminants

Microbial loads on the fruit surface were determined 6 days after treatment following the protocol described in Mehra et al. [35]. One 50-g fruit sample (~30 berries) per replicate was placed in a 0.5-L flask with 50 mL of sterile phosphate buffer (pH 7.2), and the flask was agitated on a wrist action shaker (Burrell, Pittsburg, PA, USA) at medium speed for 15 min. Aliquots of the wash buffer and 1:20 and 1:100 dilutions were plated in triplicate onto plate count agar (PCA), dichloran rose bengal chloramphenicol agar (DRBC), and Petrifilms (3M Microbiology, St. Paul, MN, USA) for enumeration of aerobic bacteria, total yeasts and molds, and *E. coli* and coliforms, respectively. PCA and DRBC dishes were incubated at room temperature and evaluated after 3 and 5 days, respectively. Petrifilms were incubated at 35 °C and colony counts made after 2 days. Colony-forming units (CFU) per gram of fruit were log-transformed and subjected to one-way analysis of variance using PROC GLM in SAS version 9.4 (SAS Institute, Cary, NC, UAS) followed by means separation using Tukey's test.

2.4 Assessment of Postharvest Disease

An initial postharvest disease assessment was made on the unshipped control following 4 days of storage at room temperature (23–25 °C) to allow latent infections to manifest themselves [35]. Subsequently, on fruit subjected to ECP™ treatment, fruit samples (60 berries per replicate) were removed from postharvest storage 6 days (trials 1 and 2) and 13 days (trial 1 only) after treatment, and similarly incubated at room temperature for 4 days. The 13-day assessment was not included in trial 2 as poor fruit quality of 'Rebel' in that trial resulted in near 100% decay after cold storage and subsequent room temperature incubation. For each assessment date and replicate,

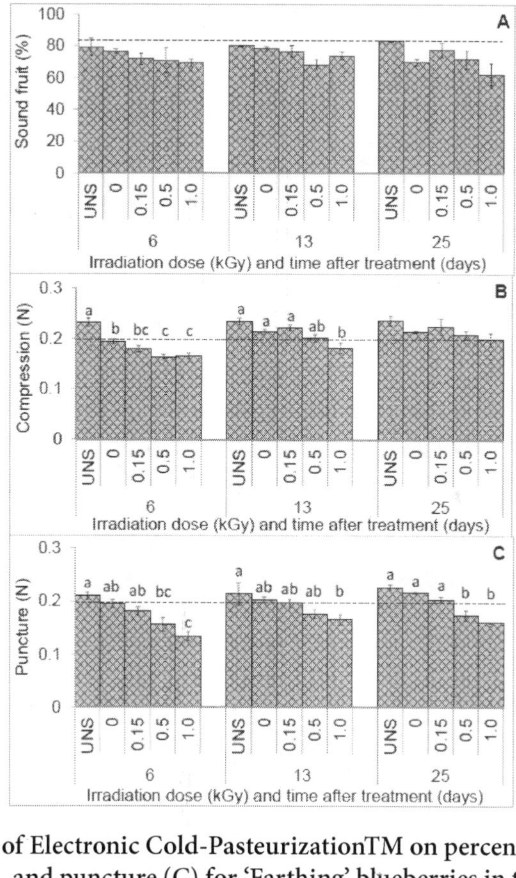

Figure 1. Effect of Electronic Cold-PasteurizationTM on percent sound fruit (A), compression (B), and puncture (C) for 'Farthing' blueberries in trial 1. Treatments included an unshipped control (UNS; not shipped to the irradiation facility) and four levels of irradiation; no irradiation control (0), 0.15, 0.5, and 1.0 kGy. Evaluations were conducted 6, 13, and 25 days after irradiation treatment. Fruit were stored at 2 to 4°C under high relative humidity until assessments were performed. An initial fruit quality assessment was performed after harvest shown as a horizontal dashed line. Means within the same storage times after treatment followed by the same letter are not significantly different from each other based on one-way analysis of variance (α = 0.05).

the number of fruit with symptoms and signs of postharvest decay was counted following examination of fruit samples with a stereo microscope. Fungal pathogens associated with diseased fruit were identified macroscopically and microscopically (utilizing both stereo- and compound microscopes) based on characteristic symptoms and signs [36,37]. Based on the number of fruit with disease symptoms and pathogen signs, postharvest disease incidence was calculated and arcsine-square root transformed for analysis by one-way analysis of variance using PROC GLM followed by means separation using Tukey's test.

3. Results

3.1 Fruit Visual Quality and Texture

To determine the effect of ECP™ treatment on fruit quality and texture, qualitative visual assessment to determine percent sound fruit, and quantitative measurements on fruit compression and skin puncture were performed (Figures 1–3). Since fruit were shipped from the site of harvest in Alma, GA, to the irradiation facility in Pocatello, ID, an unshipped control was included along with the shipped but untreated control (shipped to the treatment facility but receiving 0 kGy irradiation) to compare changes in fruit quality associated with shipping. In general, shipping did not affect fruit visual quality and texture characteristics (Figures 1–3). There were no significant effects of ECP™ on visual quality in 'Farthing' in both trials compared with the control (Figures 1A and 2A).

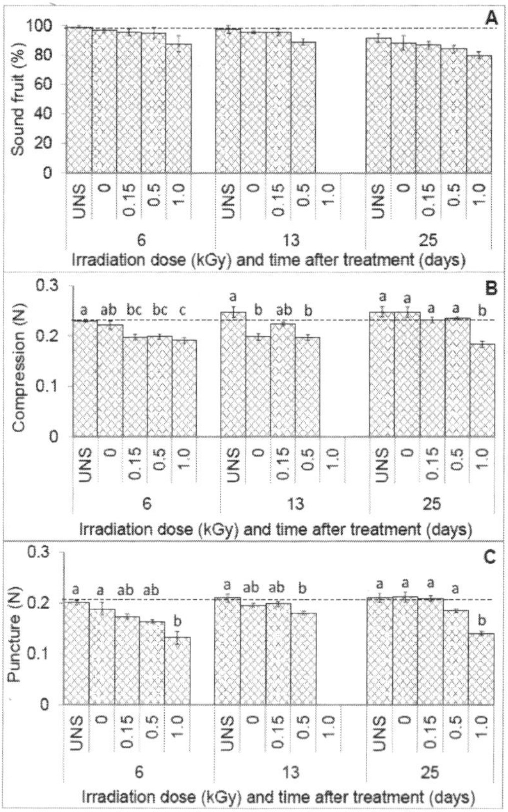

Figure 2. Effect of Electronic Cold-PasteurizationTM on percent sound fruit (A), compression (B), and puncture (C) for 'Farthing' blueberries in trial 2. Treatments included an unshipped control (UNS; not shipped to the irradiation facility) and four levels of irradiation; no irradiation control (0), 0.15, 0.5, and 1.0 kGy. Evaluations were conducted 6, 13, and 25 days after irradiation treatment. Fruit were stored at 2 to 4°C under high relative humidity until assessments were performed. An initial fruit quality assessment was performed after harvest shown as a horizontal dashed line. Due to low number of fruit, measurements were not performed for fruit treated at 1 kGy at 13 days after treatment. Means within the same storage times after treatment followed by the same letter are not significantly different from each other based on one-way analysis of variance (α = 0.05).

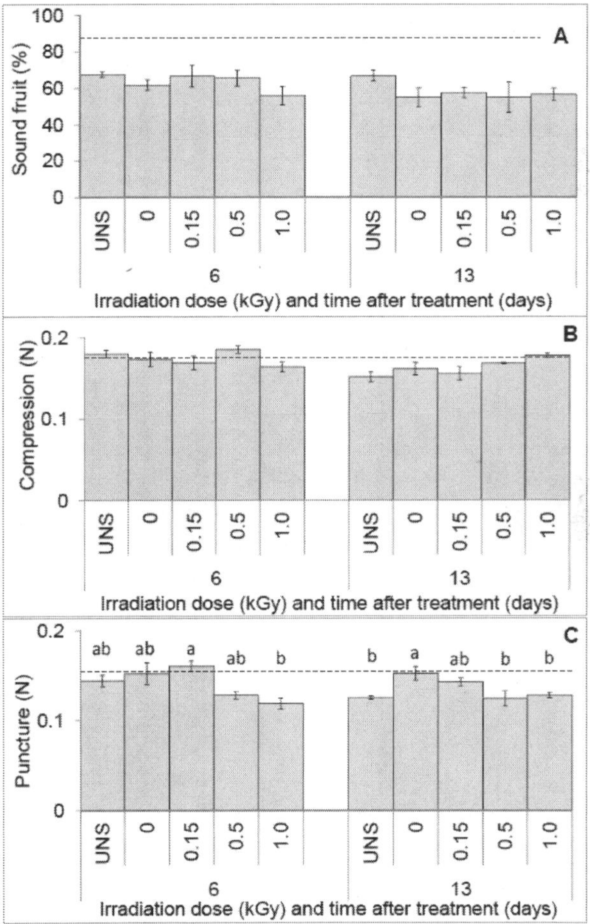

Figure 3. Effect of Electronic Cold-PasteurizationTM on percent sound fruit (A), compression (B), and puncture (C) for 'Rebel' blueberries in trial 2. Treatments included an unshipped control (UNS; not shipped to the irradiation facility) and four levels of irradiation; no irradiation control (0), 0.15, 0.5, and 1.0 kGy. Evaluations were conducted 6 and 13 days after irradiation treatment. Fruit were stored at 2 to 4

°C under high relative humidity until assessments were performed. An initial fruit quality assessment was performed after harvest shown as a horizontal dashed line. Means within the same storage times after treatment followed by the same letter are not significantly different from each other based on one-way analysis of variance (α = 0.05).

Fruit texture, measured using compression, indicated that a higher dose of irradiation resulted in a loss of firmness in 'Farthing' in both trials at various times after treatment (Figures 1B and 2B). Compared with unshipped and 0-kGy controls, a decrease in firmness was small but statistically significant with the 1.0-kGy treatment. Compared with the 0-kGy control, there was a 0.03 N decrease in firmness in the 1.0-kGy treatment at 6 and 13 days after treatment in trial 1; trial 2 showed a 0.03 N and 0.06 N at 6 and 25 days after treatment, respectively. Similarly, irradiation at 1.0 kGy resulted in a decrease in skin toughness, measured by the skin puncture force, relative to the controls in 'Farthing' in both trials. Compared with the 0-kGy control, there was a 0.04 to 0.06 N decrease in skin puncture force in the 1.0-kGy treatment at 6, 13, and 25 days after storage; trial 2 showed a 0.07 N decrease in skin puncture force at 6 and 25 days after treatment. (Figures 1C and 2C).

Comparison of fruit texture between varieties in the unshipped control at the initial and later time-points during postharvest storage indicated that 'Rebel' exhibited lower firmness and skin puncture force than 'Farthing'. 'Rebel' fruit could not be evaluated for postharvest quality attributes at 25 days after treatment due to poor quality. Visual quality of ECP™ -treated fruit of 'Rebel' did not differ from that in the control treatments (Figure 3A). There were no significant differences in fruit compression among treatments at both time points during postharvest storage (Figure 3B). Skin toughness was not different among treatments at 6 days after irradiation (Figure 3C). At 13 days after treatment, fruit irradiated at 0.5 and 1.0 kGy had lower values than the 0-kGy control, but were not different from the unshipped control suggesting that skin toughness did not change due to ECP™ in 'Rebel'.

3.2 Total Soluble Solids Content, Titratable Acidity, and Weight

There were no effects of irradiation on total soluble solids content or titratable acidity in 'Farthing' and 'Rebel' at various times after treatment (Table 1). In general, fruit weight did not change during postharvest storage. Similarly, no significant change in fruit weight was observed at various times after irradiation treatment compared with both unshipped and the 0-kGy controls in 'Farthing' and 'Rebel' (Table 2).

Table 1. Total soluble solids (TSS) content and titratable acidity (TA) of 'Farthing' and 'Rebel' blueberry fruit subjected to Electronic Cold-PasteurizationTM followed by different cold storage periods.

		Farthing Trial 1		Farthing Trial 2		Rebel Trial 2	
Days after Treatment	Treatment [a]	TSS	TA	TSS	TA	TSS	TA
	(kGy)	(% Brix)	(% CA)	(% Brix)	(% CA)	(% Brix)	(% CA)
0	UNS	13.0	0.64	13.0	0.68	8.3	0.21
6	UNS	12.4	0.51	12.2	0.56	7.9	0.20
	0	12.6	0.59	12.0	0.56	8.2	0.20

	0.15	12.7	0.54	13.0	0.45	8.2	0.23
	0.5	12.9	0.51	12.2	0.47	8.4	0.20
	1.0	13.0	0.51	12.1	0.53	8.0	0.21
13	UNS	12.6	0.51	12.8	0.53	8.0	0.20
	0	12.7	0.57	12.9	0.53	8.0	0.21
	0.15	12.8	0.54	12.0	0.51	8.1	0.21
	0.5	12.7	0.52	12.3	0.53	8.1	0.20
	1.0	12.9	0.51	-	-	8.1	0.21
25	UNS	12.8	0.45	13.0	0.38	-	-
	0	12.8	0.47	12.4	0.41	-	-
	0.15	12.7	0.47	12.8	0.42	-	-
	0.5	12.6	0.48	12.4	0.33	-	-
	1.0	12.9	0.44	12.6	0.30	-	-

a Treatments included an unshipped control (UNS, not shipped to the irradiation facility) and four levels of irradiation; no irradiation control (0), 0.15, 0.5, and 1.0 kGy. Fruit were stored at 2 to 4°C under high relative humidity until TSS and TA measurements were performed. An initial fruit quality assessment was done after harvest (day 0). Due to low number of 'Farthing' fruit in trial 2, no assessment was performed at 13 days after irradiation for fruit treated at 1.0 kGy. In case of Rebel, almost 100% decay in fruit resulted in no assessment at 25 days after treatment. One-way analysis of variance indicated no significant differences among irradiation levels within a given storage period after treatment in each trial (α = 0.05).

Table 2. Weight of 'Farthing' and 'Rebel' blueberry fruit subjected to Electronic Cold-PasteurizationTM followed by different cold storage periods.

Days after Treatment	Treatment[a] (kGy)	Farthing Trial 1	Farthing Trial 2	Rebel Trial 2
		Weight (g)	Weight (g)	Weight (g)
0	UNS	1.8	2.1	1.6
6	UNS	1.8	1.8 b	1.7
	0	1.9	2.1 ab	1.6
	0.15	2.0	2.3 a	1.7
	0.5	1.9	2.1 ab	1.7
	1.0	2.0	2.1 ab	1.7
	Prob > F	ns	0.0484	Ns
13	UNS	2.1 a	2.0	1.6
	0	1.9 ab	2.0	1.6
	0.15	1.9 ab	2.0	1.6
	0.5	1.9 ab	2.0	1.6
	1.0	1.8 b	-	1.6
	Prob > F	0.0694	ns	Ns
25	UNS	1.7	2.1	-
	0	1.8	1.9	-
	0.15	1.8	2.1	-
	0.5	1.7	1.9	-
	1.0	1.8	2.2	-
	Prob > F	ns	ns	

a Treatments included an unshipped control (UNS, not shipped to the irradiation facility) and four levels of irradiation; no irradiation control (0), 0.15, 0.5, and 1.0 kGy. Fruit were stored at 2 to 4°C under high relative humidity until weight measurements were performed. An initial fruit quality assessment was done after harvest (day 0). Due to low number of 'Farthing' fruit in trial 2, no assessment was performed at 13 days after irradiation for fruit treated at 1.0 kGy. In case of Rebel, almost 100% decay in fruit resulted in no assessment at 25 days after treatment. For every trial, means within the same storage times after treatment followed by the same letter are not significantly different from each other based on one-way analysis of variance (α = 0.05). Nonsignificant values are denoted by ns.

3.3 Microbial Load on Fruit after Treatment

Microbial loads on the fruit surface were determined for samples collected 6 days after ECP™ treatment. Microbial population densities were highest for total aerobic bacteria and total yeasts (up to ~10^5 CFU/g of fruit), followed by total molds; colony counts were lowest for coliforms (Table 3). Only a single sample of 'Rebel' showed presence of *E. coli* (at 2.7 CFU/g fruit), and therefore no statistical analysis was possible for *E. coli*. Microbial loads were similar across the two trials of 'Farthing', but were considerably higher for 'Rebel', which had very soft fruit and also the highest microbial counts (Table 3).

ECP™ significantly reduced total aerobic bacteria (by between 0.5 and 1 log units) in each of the three cultivar × trial combinations, but typically only at the 1.0-kGy irradiation level (Table 3). Yeast counts were similarly reduced in all cases, but again significant only for the 1.0-kGy level. Total surface mold counts were not reduced by irradiation in any of the cases. Population densities of coliform bacteria were not impacted on 'Farthing', but were reduced significantly and by over 2 log units on 'Rebel' (Table 3), which had the highest microbial loads in general.

Table 3. Surface microbial load, in log (colony-forming units per g of fruit), on 'Farthing' and 'Rebel' blueberry fruit subjected to Electronic Cold-PasteurizationTM 6 days after treatment.

Treatment [a] (kGy)	Aerobic Bacteria	Yeasts	Molds	Coliforms
Farthing Trial 1				
UNS	3.83 a	4.09 a	1.82	1.15
0	3.94 a	3.99 a	1.07	0.89
0.15	3.83 a	4.00 a	1.11	0.88
0.5	3.59 ab	3.77 ab	1.26	0.52
1.0	3.14 b	3.48 b	1.41	0.19
Prob > F	0.0226	0.0119	ns	ns
Farthing Trial 2				
UNS	3.15 a	3.15 a	2.31	0.73
0	3.00 a	2.98 a	1.43	0.41
0.15	3.25 a	3.21 a	1.78	0.47
0.5	3.06 a	3.03 a	1.44	0.34
1.0	2.34 b	2.47 b	1.94	0.06
Prob > F	0.0182	0.0169	ns	ns
Rebel Trial 2				
UNS	5.03 a	4.75 a	4.28	2.90 a
0	4.82 a	4.74 a	3.98	3.05 a
0.15	4.28 b	4.10 b	3.85	2.43 ab
0.5	4.10 b	4.20 ab	3.70	1.40 bc

1.0	3.93 b	3.96 b	3.38	0.85 c
Prob > F	0.0003	0.0295	ns	0.0106

a Treatments included an unshipped control (UNS, not shipped to the irradiation facility) and four levels of irradiation; no irradiation control (0), 0.15, 0.5, and 1.0 kGy. Fruit were stored at 2 to 4°C under high relative humidity until wash platings were performed. Means within the same trial and column followed by the same letter are not significantly different from each other based on one-way analysis of variance (α = 0.05). Nonsignificant values are denoted by ns.

3.4 Postharvest Disease Incidence on Fruit after Treatment

Postharvest disease incidence was determined at 6 and 13 days after treatment in trial 1 and for the 6-day post-treatment period in trial 2, each followed by a 4-day fruit exposure at room temperature to allow infections to develop. Anthracnose (caused by *Colletotrichum acutatum*), *Botrytis cinerea*, *Alternaria* sp., *Aureobasidium pullulans*, *Phomopsis vaccinii*, and *Cladosporium* sp. were observed on postharvest fruit; no significant effects of ECP™ on disease incidence were observed, neither at low decay incidence levels (<5% as observed with 'Farthing'), nor at high levels (~15% as observed with 'Rebel') (Table 4).

Table 4. Postharvest disease incidence, in percent, on 'Farthing' and 'Rebel' blueberry fruit subjected to Electronic Cold-PasteurizationTM 6 or 13 days after treatment plus 4 days at room temperature.

Days after Harvest	Treatment [a] (kGy)	Farthing Trial 1	Farthing Trial 2	Rebel Trial 2
0	UNS	0.75	1.4	17.5
6	UNS	0.83	0.42	28.3
	0	4.2	1.3	17.1
	0.15	4.6	0.63	15.4
	0.5	4.6	0.42	16.7
	1	4.2	0	14.2
	Prob > F	ns	ns	ns
13	UNS	0 b	-	-
	0	4.2 a	-	-
	150	3.8 a	-	-
	500	5.4 a	-	-
	1000	2.8 a	-	-
	Prob > F	0.011		

a Treatments included an unshipped control (UNS, not shipped to the irradiation facility) and four levels of irradiation; no irradiation control (0), 0.15, 0.5, and 1.0 kGy. Fruit were stored at 2 to 4°C under high relative humidity, followed by 4 days at room temperature, until disease assessments were performed. An initial fruit quality assessment was done after harvest (day 0). The 13-day assessment was not included in trial 2 due to low number of fruit in Farthing and nearly 100% decay in 'Rebel'. Means within the same trial and column followed by the same letter are not significantly different from each other based on one-way analysis of variance (α = 0.05). Nonsignificant values are denoted by ns.

4. Discussion

The objective of this study was to determine the effect of ECP™ on fruit quality attributes, surface microbial load, and postharvest diseases on two southern highbush cultivars. ECP™ treatment resulted in a cultivar-specific response on fruit quality. In 'Rebel',

ECP™ had no effect on visual appearance, fruit firmness, and skin toughness. In 'Farthing', however, ECP™ at 1.0 kGy, resulted in a reduction in fruit firmness and skin toughness but did not affect the visual appearance of the fruit, which was assessed based on the presence of bruises and defects such as leakiness or dents. The differential cultivar response to irradiation could be due to inherent differences in fruit firmness between the two cultivars. 'Rebel' was softer and had lower firmness and skin puncture force than 'Farthing'. Thus, irradiation may not have decreased firmness further in 'Rebel'. Similar results with differences in responses of blueberry cultivars varying in fruit texture have been observed using previous irradiation studies with various radiation sources [21,27,38]. Cultivars with firmer texture were softened after irradiation, whereas the effect of irradiation on two softer-textured cultivars varied; irradiation further softened fruit of one of the cultivars but had no effect on the other [38]. These data indicate that fruit having inherently firmer texture may be softened by irradiation, whereas the texture of fruit with lower fruit firmness may not be affected.

In this study, fruit softening and a decrease in skin toughness in 'Farthing' occurred only at the highest irradiation dose of 1.0 kGy. These results are consistent with other studies that report a dose-dependent response to irradiation with higher doses resulting in a decrease in firmness in blueberry fruit regardless of the method of irradiation. When conventional electron beam irradiation was used to treat blueberries, doses of 1.1 kGy and higher affected fruit texture resulted in softening [28]. Other studies using gamma irradiation around 0.75 kGy and higher reported increased softening in blueberries [21,27,39]. The effect of higher doses of irradiation on fruit softening has also been observed with other fruits such as raspberries [40], peaches [23,41,42], apricots [23], and grapes [43].

In spite of changes in fruit firmness, irradiation did not change other fruit quality attributes such as total soluble solids content, titratable acidity, and weight. Apart from a few minor differences, our results are consistent with other studies that indicate no effect of irradiation on fruit quality characteristics related to flavor [21,27,28,38]. The overall effect of irradiation on fruit firmness and quality in terms of consumer acceptability is an important consideration. In this study we did not perform sensory evaluations; only few other studies have conducted post-irradiation sensory analyses, and have shown mixed results related to irradiation induced softening and consumer acceptability [21,28,42] in peaches and blueberries.

In addition to fruit quality attributes, it is important to understand the effect of irradiation on the presence of fruit surface organisms that may cause foodborne illness. Blueberries are produced in open fields and can harbor various human pathogens by route of animal waste, irrigation water, and handling by farm workers. After harvest, blueberries for the fresh market are not washed nor treated for surface pathogens [20,44]. Therefore, it would be an added benefit if irradiation could reduce or eliminate such surface organisms. ECP™ treatment was effective in reducing surface microbial load in both 'Rebel' and 'Farthing'. In 'Rebel' irradiation at smaller doses was more effective in reducing surface pathogen load than in 'Farthing'. This was likely because 'Rebel' harbored a higher load of microbes on the fruit surface than 'Farthing'. In 'Rebel', aerobic bacteria and yeasts were reduced by 0.6–0.7 log units and coliforms by 2 log units at 1.0 kGy irradiation.

In 'Farthing', similar reductions were observed for aerobic bacteria and yeasts, but not for coliforms. These results are partially consistent with previous studies suggesting irradiation doses between 0.2–0.8 kGy are sufficient to cause a 1-log reduction in surface bacterial pathogens such as *E. coli* 0157:H7, *Salmonella*, and *Listeria* [32,33]. In another study with blueberries, 0.4-kGy irradiation resulted in a 1-log reduction in *Salmonella* and *Listeria* [34], but those specific taxa were not investigated in the present study. The authors concluded, and we concur, that this level of reduction may reduce risk but not guarantee safety.

Blueberries are affected by various postharvest diseases caused mainly by plant-pathogenic fungi [21,45,46]. In this study, some of the common postharvest pathogens *B. cinerea*, *Alternaria* spp., *Colletotrichum* spp., as well as *Aurebasidium*, *Phomopsis*, and *Cladosporium* were identified after postharvest storage. However, in our study ECP™ treatment did not affect the incidence of symptoms and signs associated with postharvest pathogens. Compared with microbes located on the fruit surface, a much higher dose of irradiation, typically at 1–3 kGy, is necessary to eliminate plant-pathogenic fungi [32]. Further, sensitivity of irradiation also can differ among various plant pathogens. Using an in vitro assay, inactivation of *B. cinerea*, *Penicillium expansum*, and *Rhizopus stolonifer* was observed at irradiation doses of 3–4 kGy and 1–2 kGy, respectively [47]. The maximum dose of irradiation of 1.0 kGy in our study may not have been sufficient to decrease postharvest decay pathogens. In addition, 'Farthing' had an inherently low prevalence of postharvest pathogens; hence, irradiation did not further reduce postharvest disease incidence.

Data from this study with the new ECP™ approach is in agreement with previous research which recommends a dose between 0.5 and 1.0 kGy for blueberry fruit to avoid undesirable effects on fruit quality [21,28]. While irradiation at this dose may provide protection from insect pests (not tested in this study) and some reduction in surface microbial load, more research is needed on its potential to reduce postharvest rots. In apples, mangoes, peaches, and carrots, irradiation combined with other postharvest treatments, such as cold, heat, fungicides, $CaCl_2$ treatment, or modified atmosphere offered greater benefits in controlling postharvest diseases and maintaining higher fruit quality [48–52]. Importantly, the above studies demonstrate that lower doses of irradiation are more effective when used in combination with other treatments than using irradiation alone. Blueberries are generally not treated after harvest, therefore future studies should focus on preharvest applications such as fungicides or calcium treatments in combination with irradiation and storage with modified atmosphere.

ECP™ is attractive because the method's high dose rates allow the desired irradiation dose to be obtained in a considerably shorter period of time, reducing treatment bottlenecks during operation and potentially improving produce quality through shorter treatment times outside of the cold-chain. However, direct side-by-side comparisons of ECP™ with gamma rays or X-rays at identical irradiation doses (but varying dose rates as dictated by the method) have not been conducted previously, pointing to an important research need. Future research also should address one of the limitations of our study, the need to ship the fruit to and from the treatment facility after harvest and before postharvest storage, which could have impacted treatment efficacy.

References

1. Neto, C.C. Cranberry and blueberry: Evidence for protective effects against cancer and vascular diseases. *Mol. Nutr. Food. Res.* **2007**, *51*, 652–664. [CrossRef] [PubMed]
2. Basu, A.; Rhone, M.; Lyons, T.J. Berries: Emerging impact on cardiovascular health. *Nutr. Rev.* **2010**, *68*, 168–177. [CrossRef] [PubMed]
3. Lang, G.A. Southern highbush blueberries: Physiological and cultural factors important for optimal cropping of these complex hybrids. *Acta Hortic.* **1993**, *346*, 72–80. [CrossRef]
4. Rowland, L.J.; Ogden, E.L.; Bassil, N.; Buck, E.J.; McCallum, S.; Graham, J.; Brown, A.; Wiedow, C.; Campbell, A.M.; Haynes, K.G.; et al. Construction of a genetic linkage map of an interspecific diploid blueberry population and identification of QTL for chilling requirement and cold hardiness. *Mol. Breed.* **2014**, *34*, 2033–2048. [CrossRef]
5. Evans, E.A.; Ballen, F.H. *An Overview of US Blueberry Production, Trade, and Consumption, with Special Reference to Florida*; Publication FE952; University of Florida, Institute of Food and Agricultural Sciences: Gainesville, FL, USA, 2014. Available online: http://edis.ifas.ufl.edu/fe952 (accessed on 28 August 2018).
6. United States Department of Agriculture-National Agricultural Statistics Service (USDA-NASS). *Noncitrus Fruits and Nuts, 2016 Summary*; USDA-NASS: Washington, DC, USA, 2017. Available online: http://www. clientadvisoryservices.com/Downloads/NoncFruiNu-01-23-2015.pdf (accessed on 28 August 2018).
7. Marzolo, G. *Ag Marketing Resource Center*; Iowa State University: Ames, IA, USA, 2015. Available online: http://www.agmrc.org/commodities-products/fruits/blueberries/ (accessed on 28 August 2018).
8. United States Department of Agriculture Economic Research Service (USDA-ERS). *Fruit and Tree Nut Data*; USDA-ERS: Washington, DC, USA, 2017. Available online: http://www.ers.usda.gov/data-products/fruit- and-tree-nut-data/data-by-commodity.aspx (accessed on 28 August 2018).
9. Abugoch, L.; Tapia, C.; Plasencia, D.; Pastor, A.; Castro-Mandujano, O.; López, L.; Escalona, V.H. Shelf-life of fresh blueberries coated with quinoa protein/chitosan/sunflower oil edible film. *J. Sci. Food Agric.* **2016**, *96*, 619–626. [CrossRef] [PubMed]
10. Sun, X.; Narciso, J.; Wang, Z.; Ference, C.; Bai, J.; Zhou, K. Effects of chitosan-essential oil coatings on safety and quality of fresh blueberries. *J. Food Sci.* **2014**, *79*, M955–M960. [CrossRef] [PubMed]
11. Paniagua, A.C.; East, A.R.; Hindmarsh, J.P.; Heyes, J.A. Moisture loss is the major cause of firmness change during postharvest storage of blueberry. *Postharvest Biol. Technol.* **2013**, *79*, 13–19. [CrossRef]

12. Cappellini, R.A.; Ceponis, M.J.; Koslow, G. Nature and extent of losses in consumer-grade samples of blueberries in greater New York. *HortScience* **1982**, *17*, 55–56.

13. Daykin, M.E.; Milholland, R.D. Infection of blueberry fruit by *Colletotrichum gloeosporioides*. *Plant Dis.* **1984**, *68*, 948–950. [CrossRef]

14. Smith, B.J.; Magee, J.B.; Gupton, C.L. Susceptibility of rabbiteye blueberry cultivars to postharvest diseases. *Plant Dis.* **1996**, *80*, 215–218. [CrossRef]

15. Schilder, A.M.C.; Gillett, J.M.; Woodworth, J.A. The kaleidoscopic nature of blueberry fruit rots. *Acta Hortic.* **2002**, *574*, 81–83. [CrossRef]

16. Fan, X.; Niemira, B.; Prakash, A. Irradiation of fruits and vegetables. *Food Technol.* **2008**, *3*, 37–43.

17. Palumbo, M.; Harris, L.J.; Danyluk, M.D. *Survival of Foodborne Pathogens on Berries*; Publication FSHN13-12; University of Florida, Institute of Food and Agricultural Sciences: Gainesville, FL, USA, 2013. Available online: https://edis.ifas.ufl.edu/pdffiles/FS/FS23600.pdf (accessed on 28 August 2018).

18. United States Department of Health and Human Services, Centers for Disease Control and Prevention (US CDC). *Foodborne Outbreaks*; CDC: Atlanta, GA, USA, 2016. Available online: http://wwwn.cdc.gov/ foodborneoutbreaks/Default.aspx (accessed on 28 August 2018).

19. Popa, I.; Hanson, E.J.; Todd, E.C.D.; Schilder, A.C.; Ryser, E.T. Efficacy of chlorine dioxide gas sachets for enhancing the microbiological quality and safety of blueberries. *J. Food Prot.* **2007**, *70*, 2084–2088. [CrossRef] [PubMed]

20. Gomez-Rodas, A.; Bourquin, L.; Garcia-Salazar, C.; Valera-Gomez, A.; Wise, J. *Good Agricultural Practices for Food Safety in Blueberry Production: Basic Principles*; Michigan State University Extension: East Lansing, MI, USA, 2009. Available online: http://caff.org/wp-content/uploads/2012/06/MI-Blueberry-GAPs-manual. pdf (accessed on 28 August 2018).

21. Miller, W.R.; Mitcham, E.J.; McDonald, R.E.; King, J.R. Postharvest storage quality of gamma-irradiated 'Climax' rabbiteye blueberries. *HortScience* **1994**, *29*, 98–101.

22. United States Department of Agriculture Animal and Plant Health Inspection Service (USDA APHIS). *Plant Protection and Quarantine: Treatment Manual*; USDA-APHIS: Washington, DC, USA, 2015. Available online: http://www.aphis.usda.gov/import_export/plants/manuals/ports/downloads/treatment.pdf (accessed on 28 August 2018).

23. Drake, S.R.; Neven, L.G. Irradiation as an alternative to methyl bromide for quarantine treatment of stone fruits. *J. Food Qual.* **1998**, *22*, 529–538. [CrossRef]

24. Aegerter, A.F.; Folwell, R.J. Economic aspects of alternatives to methyl bromide in the postharvest and quarantine treatment of selected fresh fruits. *Crop Prot.* **2000**, *19*, 161–168. [CrossRef]

25. Kitinoja, L. *Use of Cold Chains for Reducing Food Losses in Developing Countries*; White Paper 13-03; The Postharvest Education Foundation (PEF): La Pine, OR, USA, 2013. Available online: http://www. postharvest.org/use%20of%20cold%20chains%20 pef%20white%20paper%2013-03%20final.pdf (accessed on 28 August 2018).

26. Morehouse, K.M.; Komolprasert, V. Irradiation of Food and Packaging: An Overview. In *Irradiation of Food and Packaging*; ACS Symposium Series 875; American Chemical Society: Washington, DC, USA, 2004. Available online: https://www.fda.gov/food/ ingredientspackaginglabeling/irradiatedfoodpackaging/ ucm081050.htm (accessed on 28 August 2018).

27. Miller, W.R.; McDonald, R.E.; Smittle, B.J. Quality of 'Sharpblue' blueberries after electron beam irradiation. *HortScience* **1995**, *30*, 306–308.

28. Moreno, M.A.; Castell-Perez, M.E.; Gomes, C.; Da Silva, P.F.; Moreira, R.G. Quality of electron beam irradiation of blueberries (*Vaccinium corymbosum* L.) at medium dose levels (1.0–3.2 kGy). *Food Sci. Technol.* **2007**, *40*, 1123–1132.

29. Golding, J.B.; Blades, B.L.; Satyan, S.; Jessup, A.J.; Spohr, L.J.; Harris, A.M.; Banos, C.; Davies, J.B. Low dose gamma irradiation does not affect the quality, proximate or nutritional profile of 'Brigitta' blueberry and 'Maravilla' raspberry fruit. *Postharvest Biol. Technol.* **2014**, *96*, 49–52. [CrossRef]

30. Lires, C.M.L.; Docters, A.; Horak, C.I. Evaluation of the quality and shelf life of gamma irradiated blueberries by quarantine purposes. *Radiat. Phys. Chem.* **2018**, *143*, 79–84. [CrossRef]

31. United States Food and Drug Administration (US FDA). *Code of Federal Regulations Title 21*; US FDA: Washington, DC, USA, 2016. Available online: https://www. accessdata.fda.gov/scripts/cdrh/cfdocs/cfcfr/ cfrsearch.cfm?fr=179.26 (accessed on 28 August 2018).

32. Niemira, B.A.; Sommers, C.H. New applications in food irradiation. In *Encyclopedia of Agricultural, Food and Biological Engineering*, 1st ed.; Heldman, D.R., Ed.; Taylor & Francis: New York, NY, USA, 2006; pp. 1–6.

33. Niemira, B.A.; Fan, X. Irradiation enhances quality and safety of fresh and fresh-cut fruits and vegetables. In *Microbial Safety of Fresh Produce*; Niemira, B.A., Doona, C.J., Feeherry, F.E., Gravani, R.B., Eds.; Wiley-Blackwell: Ames, IA, USA, 2009; pp. 191–204. ISBN 9781444319347.

34. Thang, K.; Au, K.; Rakovski, C.; Prakash, A. Effect of phytosanitary irradiation and methyl bromide fumigation on the physical, sensory, and microbiological quality of blueberries and sweet cherries. *J. Sci. Food Agric.* **2016**, *96*, 4382–4389. [CrossRef] [PubMed]

35. Mehra, L.K.; MacLean, D.D.; Savelle, A.T.; Scherm, H. Postharvest disease development on southern highbush blueberry fruit in relation to berry flesh type and harvest method. *Plant Dis.* **2013**, *97*, 213–221. [CrossRef]

36. Barnett, H.L.; Hunter, B.B. *Illustrated Genera of Imperfect Fungi*, 4th ed.; Macmillan: New York, NY, USA, 1987; ISBN 9780890541920.

37. Wharton, P.; Schilder, A. *Blueberry Fruit Rot Identification Guide*; E-2847; Michigan State University: East Lansing, MI, USA, 2003. Available online: http://msue.anr.msu.edu/uploads/resources/pdfs/Michigan_Blueberry_Facts_-_Blueberry_Fruit_Rot_Identification_Guide_(E2847).pdf (accessed on 28 August 2018).

38. Eaton, G.W.; Meehan, C.; Turner, N. Some physical effects of postharvest gamma radiation on the fruit of sweet cherry, blueberry, and cranberry. *J. Can. Inst. Food Sci. Technol.* **1970**, *3*, 152–156. [CrossRef]

39. Trigo, M.J.; Sousa, M.B.; Sapata, M.M.; Ferreira, A.; Curado, T.; Andrada, L.; Ferreira, E.S.; Antunes, C.; Horta, M.P.; Pereira, A.R.; et al. Quality of gamma irradiated blueberries. *Acta Hortic.* **2006**, *715*, 573–578. [CrossRef]

40. Guimarães, C.; Menezes, E.G.T.; de Abreu, P.S.; Rodrigues, A.C.; Borges, P.R.S.; Batista, L.R.; Cirilo, M.A.; Lima, L.C.O. Physicochemical and microbiological quality of raspberries (*Rubus idaeus*) treated with different doses of gamma irradiation. *Food Sci. Technol.* **2013**, *33*, 316–322. [CrossRef]

41. Kim, M.S.; Kim, K.H.; Yook, H.S. The effects of gamma irradiation on the microbiological, physicochemical and sensory quality of peach (*Prunus persica* L. Batsch Cv. Dangeumdo). *J. Korean Soc. Food Sci. Nutr.* **2009**, *38*, 364–371. [CrossRef]

42. McDonald, H.; McCulloch, M.; Caporaso, F.; Winborne, I.; Oubichon, M.; Rakovski, C.; Prakash, A. Commercial scale irradiation for insect disinfestations preserves peach quality. *Radiat. Phys. Chem.* **2012**, *81*, 697–704. [CrossRef]

43. Kim, C.G.; Rakowski, C.; Caporaso, F.; Prakash, A. Low-dose irradiation can be used as a phytosanitary treatment for fresh table grapes. *J. Food Sci.* **2014**, *79*, S81–S91. [CrossRef] [PubMed]

44. Palumbo, M.; Harris, L.J.; Danyluk, M.D. *Outbreaks of Foodborne Illness Associated with Common Berries, 1983 through May 2013*; Publication FSHN13-08; University of Florida, Institute of Food and Agricultural Sciences: Gainesville, FL, USA, 2013. Available online: http://ucfoodsafety.ucdavis.edu/files/223896.pdf (accessed on 28 August 2018).

45. Ceponis, M.J.; Capellini, R.A. Control of postharvest decays of blueberry fruits by precooling, fungicide, and modified atmospheres. *Plant Dis.* **1979**, *63*, 1049–1053.

46. Hardenburg, R.E.; Watada, A.E.; Wang, C.Y. The commercial storage of fruits, vegetables, and florist and nursery stocks. *USDA Agric. Handb.* **1986**, *66*, 240–242.

47. Jeong, R.D.; Shin, E.J.; Chu, E.H.; Park, H.J. Effects of ionizing radiation on postharvest fungal pathogens. *Plant Pathol. J.* **2015**, *31*, 176–180. [CrossRef] [PubMed]

48. Spalding, D.H.; Reeder, W.F. Decay and acceptability of mangoes treated with combinations of hot water, imazalil, and γ-radiation. *Plant Dis.* **1986**, *70*, 1149–1151. [CrossRef]

49. Johnson, G.I.; Boag, T.S.; Cooke, A.W.; Izard, M.; Panitz, M.; Sangchote, S. Interaction of post harvest disease control treatments and gamma irradiation on mangoes. *Ann. Appl. Biol.* **1990**, *116*, 245–257. [CrossRef]

50. Lacroix, M.; Lafortune, R. Combined effects of gamma irradiation and modified atmosphere packaging on bacterial resistance in grated carrots (*Daucus carota*). *Radiat. Phys. Chem.* **2004**, *71*, 79–82. [CrossRef]

51. Lafortune, R.; Caillet, S.; Lacroix, M. Combined effects of coating, modified atmosphere packaging and gamma irradiation on quality maintenance of ready-to-use carrots (*Daucus carota*). *J. Food Prot.* **2005**, *68*, 353–359. [CrossRef] [PubMed]

52. Hussain, P.R.; Dar, M.A.; Wani, A.M. Effect of edible coating and gamma irradiation on inhibition of mould growth and quality retention of strawberry during refrigerated storage. *Int. J. Food Sci. Technol.* **2012**, *47*, 2318–2324. [CrossRef]

Chapter-4

The Effect of Ethephon, Abscisic Acid, and Methyl Jasmonate on Fruit Ripening in Rabbiteye Blueberry (*Vaccinium virgatum*)

1. Introduction

Blueberries (*Vaccinium* spp.) contain bioactive compounds which offer potential health benefits and have witnessed a large increase in production over the last two decades [1,2]. Blueberries are native to North America and some common cultivated species include lowbush (*Vaccinium angustifolium* Ait.), northern highbush (*Vaccinium corymbosum* L.), rabbiteye (*V. virgatum* Ait.), and southern highbush (hybrids of *V. corymbosum, V. virgatum,* and *V. darrowii* Camp.) [3–5]. During fruit development, blueberry fruit on a branch mature at different rates, resulting in a non-uniform ripening period extending over 2 to 3 weeks [6]. As a result, blueberries intended for the fresh fruit market are hand harvested three to five times depending on the variety. This makes harvesting a labor intensive and expensive component of blueberry production, requiring up to 520 h of labor/acre and costing up to $0.70 per pound of harvested fruit [7–9]. Concentrating the period of ripening could help reduce the required number of harvests and reduce costs associated with production. Ripening is regulated by multiple plant hormones such as ethylene, abscisic acid, auxins, and jasmonates [10]. External applications of plant growth regulators (PGRs) that influence the levels or activity of these plant hormones may alter the progression of ripening and thereby help in concentrating the period of fruit ripening for efficient harvesting. Therefore, understanding the progression of ripening and developing tools such as PGR applications can help in improving the efficiency of blueberry harvesting.

Fruit ripening is a coordinated process involving changes in fruit texture, color, flavor, and susceptibility to biotic and abiotic factors [11,12]. Although all fruit display these changes during ripening, fruits can be generally classified into one of two types depending on physiological and biochemical changes accompanying the initiation and progression of ripening: climacteric and non-climacteric. In climacteric fruits such as tomato (*Solanum lycopersicum*), banana (*Musa* spp.), and apple (*Malus* × *domestica*), ripening is accompanied by a peak in respiration and ethylene production [11–14]. In such fruits, once ethylene production is triggered at ripening, it is autocatalytic and is one of the key factors that regulate changes associated with ripening. Non-climacteric fruits, such as strawberry (*Fragaria × ananassa*) and grape (*Vitis vinifera*), do not exhibit an increase in respiration and ethylene in association with ripening. In these fruits, the role of ethylene and other signals in regulating ripening are not completely understood [12,14–16]. The roles of climacteric respiration and ethylene in the progression of fruit ripening in blueberry are unclear. Some previous studies observed an increase in respiration and ethylene during blueberry ripening, suggesting a potential climacteric nature to the ripening process [6,17,18]. Also, external application of the ethylene-releasing compound ethephon accelerated the progression of ripening and reduced the harvest time in blueberry [19–21]. However, several other studies have classified blueberry as a non-climacteric fruit that does not display a substantial climacteric rise in respiration or ethylene evolution [22]. Hence, further studies are required to better understand the contribution of ethylene in blueberry ripening, and to determine if manipulation of this plant hormone offers a viable option for controlling ripening.

Abscisic acid (ABA), another plant hormone, plays an important role in many developmental processes such as adaptation to stress and seed dormancy. In addition, recent work has suggested a role for ABA during ripening in climacteric as well as non-climacteric fruit. Abscisic acid concentration increases during fruit ripening in apple [23], orange (*Citrus sinensis*) [24], cherry (*Prunus avium*) [25], strawberry [26,27], and grape [28]. In strawberry, decreased expression of *9-cis-epoxycarotenoid dioxygenase* (*FaNCED1*), a gene coding for an enzyme involved in ABA biosynthesis, lowered ABA levels and prevented fruit from ripening normally [27]. In grape, ABA applications improved red color and helped achieve early harvest, underlining its potential for accelerating ripening [29–31]. Further, in tomato, ABA may function upstream of ethylene and induce the expression of ethylene biosynthesis genes to regulate ripening [32,33]. Similarly, in banana, ABA applications may enhance ethylene sensitivity and coordinate ethylene-regulated ripening [34]. In bilberry (*V. myrtillus* L.), which is closely related to blueberry, ABA has been implicated in the regulation of ripening [35]. In highbush blueberry (*V. corymbosum*), ABA levels increase at the onset of ripening and may be involved in regulating the production of flavonoids [36]. However, external applications of ABA delayed ripening and increased fruit firmness in southern highbush blueberry (*V. corymbosum* interspecific hybrids) [37]. Although these studies suggest a potential role for ABA in regulating blueberry ripening, it requires further investigation, especially to determine if external ABA applications can be used to reliably manipulate the progression of this process across different blueberry species.

Jasmonates are another group of phytohormones with well-characterized roles in defense responses and developmental processes such as senescence [38]. Jasmonates have been implicated recently in the regulation of fruit ripening [14,39]. In tomato and apple, jasmonates promoted ethylene biosynthesis by inducing the expression of genes involved in its biosynthesis [40]. In apple, methyl jasmonate (MeJA) applications influenced the production of aromatic volatiles, an integral component of fruit flavor, in a cultivar-dependent manner [41]. In peach (*Prunus persica*), jasmonates delayed ripening [42]. Although MeJA had a negative effect on ethylene biosynthesis during ripening in peach, it still promoted anthocyanin biosynthesis [43]. In non-climacteric fruits such as cultivated strawberry and Chilean wild strawberry (*Fragaria chiloensis*), application of MeJA increased ethylene evolution and respiration, and promoted color development thereby accelerating ripening [44,45]. In raspberry (*Rubus idaeus*), MeJA application increased flavonoid content, total soluble solids (TSS) content, and total sugars, and lowered titratable acidity (TA), thus influencing multiple ripening characteristics [46]. Together, these emerging data suggest that the effect of jasmonates on fruit ripening may be species-specific, requiring further evaluation in the species of interest. Further, preharvest and postharvest applications of MeJA may not only improve fruit quality but also offer a protective role by limiting pathogen growth as seen in strawberry and peach [39]. MeJA applications on highbush blueberry resulted in changes in total sugar content, total anthocyanin content, and expression of anthocyanin biosynthesis genes in a cultivar-dependent manner [47]. However, the specific role of MeJA in blueberry ripening and its effect on postharvest fruit quality attributes is not clear and has not been investigated previously.

While the effects of multiple PGRs on fruit ripening have been evaluated in various fruit crops, these have not yet been tested extensively in blueberry. Blueberry production could greatly benefit from the use of PGRs that help manipulate the time of ripening. Hence, the main goal of this research was to evaluate three PGRs, ethephon, ABA, and MeJA, for their ability to alter the progression of ripening in two rabbiteye blueberry cultivars. These three PGRs were selected for further study due to previous research suggesting their potential as described above. Furthermore, as preharvest applications of these PGRs can influence postharvest fruit quality and storage characteristics including disease symptom development, the effects of their application on postharvest fruit quality and disease incidence were also evaluated.

2. Materials and Methods

2.1 Plant Material and PGRs

Two rabbiteye blueberry cultivars, Premier and Powderblue (both at 5 years since planting), grown at the Durham Horticulture Farm in Watkinsville, GA were used for this study. All applications were performed when around 30–40% of fruit on the plant were ripe. Whole plants were sprayed using a hand-held sprayer until run-off. For the early-maturing Premier, the treatments consisted of: control (water), 250 mg L^{-1} ethephon, 600 mg L^{-1} ABA, and 0.5 mM MeJA. All treatments were applied on 20 June 2016 along with an adjuvant (0.15% Latron B-1956; Simplot, Lathrop, CA, USA). The doses were determined based on preliminary studies. Applications on Premier were made in the evening close to sunset to minimize photo-destruction of ABA. For the later-maturing Powderblue, the same treatments were applied on 9 July 2016 except that the concentration of ABA and MeJA were increased to 1000 mg L^{-1} and 1 mM, respectively. Due to potential rainfall in the late afternoon, all applications on Powderblue were made early in the morning. For each treatment, four replicates consisting of four individual plants were used in both cultivars.

2.2 Rate of Ripening

Prior to PGR application, three 50 to 100-cm-long shoots, each consisting of a total of approximately 50–100 fruit, were tagged per replicate. Very small immature as well as ripe fruit were removed from the tagged branches. The number of green, pink, and ripe fruit was counted prior to and after PGR applications at regular intervals (2–4 days) up to 11 days and 13 days for Premier and Powderblue, respectively. Fruit counted as pink ranged from having around 25% pink color (75% green) to around 75% pink (25% blue) on the fruit surface. Fruit was considered ripe when the color of the entire fruit was blue. The percentage of green, pink, and ripe fruit was calculated from these data for each assessment date.

2.3 Postharvest Fruit Quality and Disease Incidence

Two additional shoots containing around 300 fruit (total) were tagged on each replicate to study the effect of PGR applications on postharvest fruit quality and disease incidence. Very small immature and ripe fruit at the time of application were removed. Ripe fruit were hand-harvested approximately 10 days after application of PGRs and split into three groups for postharvest fruit quality analyses. These groups were randomly assigned to one of the following treatment periods for postharvest evaluation: PH + 1 (postharvest + 1 day); PH + 15 (postharvest + 15 days); and PH + 29 (postharvest + 29 days). For postharvest storage, fruit were placed in a walk-in cooler set to 4 °C and a relative humidity of 90–95%. For each storage period and replicate, around 40 fruit were used for fruit quality evaluation and around 60 fruit were used for disease incidence evaluation. For fruit quality analysis, visual evaluation of quality was conducted and weight, texture, pH, TA, TSS content, and berry color were measured. For visual evaluation of fruit quality, 30 fruit per replicate were scored for symptoms of bruising such as tears, dents, leakiness, and appearance of mold to determine the percent defective fruit. Fruit weight was measured on 20 fruit, using a balance (Quintix® Precision Balance, Sartorius, Bohemia, NY, USA). Fruit texture measurements were made using a fruit texture analyzer (GS-15, Güss Manufacturing, Strand, South Africa). Two tests, compression and skin puncture, were performed on 12 fruit per replicate for determining fruit texture by orienting the fruit on the equatorial plane. For compression analyses, a probe with a 15-mm diameter end plate was used with parameters set at a measure speed of 5 mm s^{-1} and measure distance of 1 mm. To measure skin puncture force, a 1.5-mm probe was used with parameters set at a measure speed of 5 mm s^{-1} and measure distance of 3 mm.

For measuring pH, TA, and TSS, juice from around 30 g of fruit was extracted using a blender followed by centrifugation for 10 min at 3901× g on a benchtop centrifuge (Allegra X-22, Beckman Coulter Life Sciences, Indianapolis, IN, USA). The supernatant was filtered through cheesecloth. Around 0.3 mL of supernatant was used to determine TSS using a digital handheld refractometer (Atago USA, Belleveue, WA, USA). To determine pH and TA, the supernatant was titrated using an automatic mini titrator (Hanna Instruments, Woonsocket, RI, USA) and alkaline titrant. The titrator has a pH electrode which provided an initial pH value of the supernatant before titration is initiated. For TA, the data are expressed as percent citric acid (CA) equivalents. Fruit color was determined on 20 fruit using a handheld colorimeter (3nh Technology Co., Shenzhen, China).

To determine natural postharvest disease incidence, fruits were maintained at 23–25 °C for 4 days after removing them from cold storage at the three postharvest intervals described above. Fruit displaying symptoms of disease and/or signs of plant pathogens were recorded using around 60 fruit per replicate. The associated pathogens were identified microscopically as described in Mehra et al. [48].

Statistical analysis (one-way analysis of variance for a randomized complete block design) was performed separately for every time-point after harvest using JMP Pro 12 (SAS Institute, Cary, NC, USA). Means were separated using Tukey's honestly significant difference (HSD) test (α = 0.05).

3. Results

3.1 Effect of PGR Application on Fruit Ripening

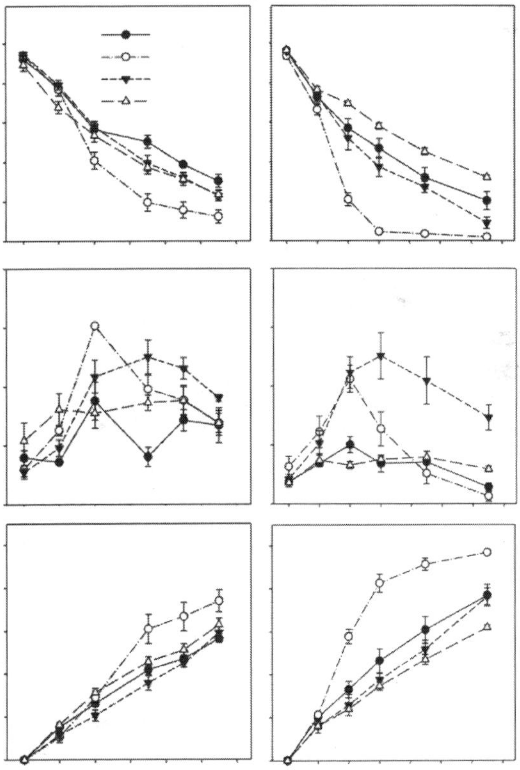

Figure 1. Effect of preharvest treatments with water (contol), ethephon, abscisic acid (ABA), and methyl jasmonate (MeJA) on ripening of rabbiteye blueberry, Premier (A–C) and Powderblue (D–F). Values are means and standard errors of four replicates. Within each assessment period, means with the same letter are not significantly different according to ANOVA and Tukey's HSD (α = 0.05).

In both cultivars, the proportion of green fruit decreased while that of ripe fruit increased over the duration of the experiment (Figure 1). In Premier, ethephon-treated fruit had a lower proportion of green fruit than that in the control from 4 days after treatment (Figure 1A). At this stage, the proportion of pink fruit was higher in ethephon-treated fruit (Figure 1B). The proportion of ripe fruit was significantly higher in the ethephon treatment from 7 days after treatment compared with the control (Figure 1C). At 7 days after treatment, 42% of the fruit were ripe in the control compared with 61% in the ethephon treatment. In contrast, treatment with ABA did not affect the proportion of green or ripe fruit but transiently increased the proportion of pink fruit at 7 days after treatment, compared withthe control (Figure 1B). Similarly, treatment with MeJA did not alter the proportion of green and ripe fruit compared with the control, but increased the proportion of pink fruit at 2 days and 7 days after treatment (Figure 1B).

In Powderblue, the doses of application of ABA and MeJA were higher (see Section 2) as these PGRs did not appear to affect ripening in Premier at lower doses of application. The PGR applications generally resulted in similar effects on Powderblue fruit ripening as in Premier, with a few exceptions (Figure 1D–F). The proportion of green fruit was lower (Figure 1D), and ripe fruit was higher (Figure 1F) than in the control in ethephon-treated fruit starting from 4 days after treatment until the end of evaluation; and, the proportion of pink fruit was higher at 4 days after treatment (Figure 1E). At 4 days after treatment, while only 33% of the fruit were ripe on the control plants, around 58% were ripe in response to the ethephon treatment (Figure 1F). Application of ABA did not affect the proportion of green or ripe fruit compared with the control but increased the proportion of pink fruit from 4 days after application (Figure 1E). At the rate of ABA used in this study (1000 mg L^{-1}), phytotoxicity symptoms were observed in leaves (data not shown). Application of MeJA did not alter the proportion of green, pink, or ripe fruit at any stage after treatment in comparison with the control (Figure 1D–F).

3.2 Effect of PGR Application on Fruit Color

None of the fruit color-related parameters were significantly different among the PGR treatments in Premier at 1 day after harvest (Table 1). In Powderblue, treatment with ethephon and ABA also did not alter any of the fruit color-related parameters with respect to the control treatment at 1 day after harvest. In response to MeJA treatment, however, the parameters L*, which measures the lightness, and b*, which measures yellow/blue color, were higher and lower respectively, indicating lighter and greater blue fruit color than in the control (Table 1).

Table 1. Effect of preharvest treatment with water (control), ethephon, abscisic acid (ABA), and methyl jasmonate (MeJA) on fruit color after 1 day of cold storage at 4°C in Premier and Powderblue blueberry.

Cultivar/Treatment Z	L *	a *	b *	c *	h *
Premier					
Control	38.0	1.1	6.3	6.4	260.0
thephon	38.3	1.0	6.3	6.4	261.2

ABA	37.9	1.0	5.9	6.1	260.2
MeJA	38.1	1.0	6.3	6.5	260.9
Significance	NS	NS	NS	NS	NS
Powderblue					
Control	40.9b	1.2	6.37a	6.6ab	260.5
Ethephon	43.3ab	1.3	6.44ab	6.6ab	258.1
ABA	40.7b	1.1	6.16a	6.3b	260.1
MeJA	44.0a	1.4	6.77b	6.9a	258.6
Significance	0.0078	NS	0.0066	0.0063	NS

z Means followed by the same letter within a column are not significantly different, according to Tukey's HSD (α = 0.05).

3.3 Effect of PGR Application on Fruit Quality during Postharvest Storage

Visual assessment of postharvest fruit quality using variables such as bruises, dents, and mold incidence during postharvest storage indicated a 20% increase in the percentage of defective fruit from 1 day until 29 days after storage in Premier (Table 2). There were no significant effects of the PGR treatments until 29 days after harvest in Premier (Table 2). At 29 days after harvest, ABA application resulted in a higher proportion whereas ethephon application resulted in a lower proportion of defective fruit compared with the control. In Powderblue the percentage of defective fruit increased by 15% from 1 day until 29 days after storage in control fruit; none of the PGR applications significantly affected the visually assessed variables for fruit quality (Table 2).

In Premier, fruit compression and puncture declined in the control by 18 and 20%, respectively, at 29 days after storage compared with 1 day after storage (Table 3). In Premier, ABA applications reduced the force required for fruit compression at 29 days of postharvest storage by ~17%, suggesting a decrease in fruit firmness relative to the control (Table 3). None of the other treatments affected fruit texture characteristics or the other fruit quality characteristics such as fruit weight, TSS, TA, and juice pH, evaluated during postharvest storage with respect to the control (Tables 3 and 4). In Powderblue, fruit compression and puncture declined in the control by 17% and 26%, respectively, at 29 days after storage compared with 1 day after storage (Table 3). In Powderblue, fruit firmness as measured by compression was higher by 16% in ethephon-treated fruit compared with the control at 15 days of postharvest storage (Table 3). Fruit weight did not differ among various treatments during postharvest storage with respect to the control (Table 3). Ethephon treatment resulted in higher TA values than that in the control at various times after storage, although this was significant only at 15 days of postharvest storage (by 21%) (Table 4). TSS was lower in the ABA treatment than in the control at 15 days after harvest by ~11%. Also, juice pH was lower in response to ethephon and MeJA treatments than in the control at 1 day after harvest (Table 4).

Table 2. Percent defective fruit determined at various times after harvest in Premier and Powderblue blueberry following preharvest treatment with water (control), ethephon, abscisic acid (ABA), and methyl jasmonate (MeJA).

		Defective Fruit (%) z	
Cultivar/Treatment	**1 Day**	**15 Days**	**29 Days**
Premier			
Control	3.3	19.2	23.3b
Ethephon	2.2	7.8	10.0c
ABA	3.3	16.7	43.3a
MeJA	11.7	15.0	28.3ab
Significance	NS	NS	0.0003
Powderblue			
Control	5.0	13.3	20.0
Ethephon	5.0	6.7	19.2
ABA	4.2	8.3	18.3
MeJA	5.8	11.7	21.7
Significance	NS	NS	NS

z Means followed by the same letter within a column for a given time-point after storage are not significantly different, according to Tukey's HSD (α = 0.05).

Table 3. Effect of preharvest treatment with water (control), ethephon, abscisic acid (ABA), and methyl jasmonate (MeJA) on fruit texture and weight sampled at 1, 15, and 29 days of cold storage at 4°C in Premier and Powderblue blueberry.

			Berry Texture z					
Compression (kgF)			Pressure (kgF)			Berry Weight (g) z		
Cultivar/Treat1mdent	**15 d**	**29 d**	**1 d**	**15 d**	**29 d**	**1 d**	**15 d**	**29 d**
Premier								
Control 0.22	0.20	0.18a	0.15	0.15	0.12	0.81ab	0.82ab	0.81
Ethephon 0.23	0.20	0.19a	0.15	0.15	0.12	0.77ab	0.80ab	0.76
ABA 0.20	0.19	0.15b	0.14	0.14	0.11	0.86a	0.88a	0.79
MeJA 0.23	0.21	0.19a	0.15	0.16	0.13	0.70b	0.72b	0.70
Significance NS	NS	0.0166	NS	NS	NS	0.0229	0.0067	NS
Powderblue								
Control 0.23	0.19b	0.19ab	0.19	0.15	0.14	0.87	0.83	0.82
Ethephon 0.26	0.22a	0.21a	0.18	0.15	0.15	0.64	0.68	0.69
ABA 0.24	0.20b	0.18b	0.18	0.15	0.14	0.84	0.83	0.89
MeJA 0.24	0.21ab	0.17b	0.19	0.16	0.15	0.80	0.77	0.78
Significance NS	0.0077	0.0120	NS	NS	NS	NS	NS	NS

z Means followed by the same letter within a column for a given time-point after storage are not significantly different, according to Tukey's HSD (α = 0.05).

Table 4. Effect of preharvest treatment with water (control), ethephon, abscisic acid (ABA), and methyl jasmonate (MeJA) on fruit quality sampled after 1, 15, and 29 days of cold storage at 4°C in Premier and Powderblue blueberry.

Total Soluble Solids (Brix) z			Titratable Acidity (%) z			Juice pH z		
Cultivar/Treat1mdent 15 d		29 d	1 d	15 d	29 d	1 d	15 d	29 d
Premier								
Control 11.2	10.6	10.6	0.40	0.34	0.27	3.48	3.60	3.70
Ethephon 9.7	9.8	9.4	0.46	0.37	0.35	3.47	3.60	3.53
ABA 10.9	9.6	9.9	0.37	0.34	0.29	3.53	3.60	3.60
MeJA 10.5	9.8	9.9	0.44	0.36	0.30	3.43	3.58	3.70
Significance NS	NS	NS	NS	NS	NS	NS	NS	NS
Powderblue Control 12.7	13.1a z	13.2	0.48	0.45b	0.36	3.48a	3.48a	3.45
Ethephon 12.0	12.1ab	12.6	0.56	0.54a	0.41	3.35b	3.38a	3.45
ABA 11.4	11.6b	12.4	0.49	0.50ab	0.36	3.40ab	3.40a	3.48
MeJA 12.1	13.0a	13.2	0.60	0.53ab	0.41	3.35b	3.38a	3.43
Significance NS	0.0166	NS	NS	0.0490	NS	0.0150	0.0486	NS

z Means followed by the same letter within a column for a given time-point after storage are not significantly different, according to Tukey's HSD (α = 0.05).

3.4 Effect of PGR Application on Postharvest Disease Incidence During Storage

The major postharvest pathogens indicated by disease symptoms and signs in this study were *Colletotrichum acutatum* (causal agent of anthracnose fruit rot), *Phomopsis vaccinii*, *Botrytis cinerea* (gray mold), *Alternaria* spp., and *Pestalotia* spp. Postharvest disease incidence in both Premier (typically < 5%) and Powderblue (typically < 10%) was low, despite the 4-day incubation period at room temperature following various postharvest storage periods. Due to low pathogen counts, only overall postharvest disease incidence was analyzed; no significant differences among treatments at different time intervals of storage were observed (Figure 2).

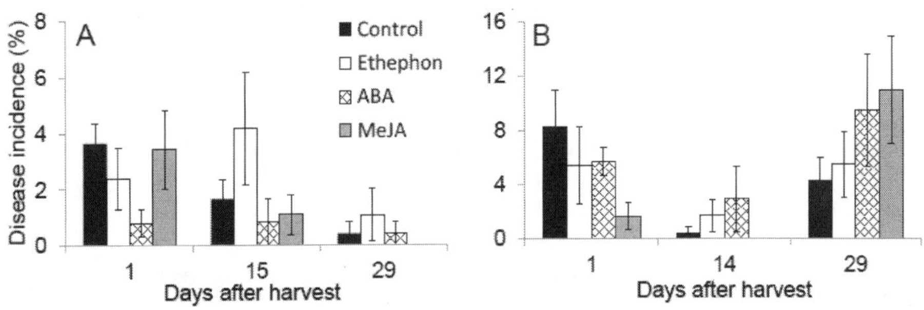

Figure 2. Percent postharvest disease incidence determined at various times after harvest in Premier (A) and Powderblue (B) rabbiteye blueberry following

applications of water (control), ethephon, abscisic acid (ABA), and methyl jasmonate (MeJA). Values are means and standard errors of four replicates and 40 to 60 fruit per replicate. No significant differences ($\alpha = 0.05$) were detected among the treatments.

4. Discussion

Data from this study clearly indicate that both of the rabbiteye blueberry cultivars responded rapidly to ethephon applications. The time taken for 50% of fruit to ripen was advanced by up to 3 days after ethephon treatment compared with the control (Figure 1). Ethephon also increased the number of ripe fruit; ripe fruit increased from 42% to 61% in Premier at 7 days after application and 46% to 83% in Powderblue 6 days after treatment, indicating that the application of this PGR can accelerate the progression of ripening and reduce the time to harvest in blueberry. Several previous studies demonstrated that ethephon accelerates ripening and can reduce the number of required harvests in blueberry [19–21,49]. Results from the current study further expand these findings of acceleration of fruit ripening by ethephon to rabbiteye blueberry.

For a PGR to be effective as a ripening aid, in addition to accelerating ripening it should display minimal negative effects on postharvest fruit quality. Ethephon treatment resulted in a substantial decrease in the proportion of defective fruit after postharvest storage (29 days), at least in Premier. However, fruit texture was not affected by ethephon application in Premier. In Powderblue, compression force at 15 days after ethephon treatment was slightly higher but was not different by 29 days after storage, indicating that ethephon had minimal and temporary effects on fruit firmness characteristics during postharvest storage. These data are generally consistent with those of Dekazos [20] who used rabbiteye blueberry. However, Ban et al. [21] reported substantial reduction in firmness in another rabbiteye blueberry cultivar, Tifblue, in response to ethephon. In that study, fruit slices were used for analysis of firmness rather than intact fruit as used in the current study as well as several others [20,50,51], which may explain the different observations. In the current study, no change in fruit weight in response to ethephon treatment was observed, consistent with results from a study on Tifblue treated with 200 mg L^{-1} ethephon [21], and in two highbush blueberry genotypes treated with similar doses of ethephon (240 mg L^{-1}) [19]. With similar or comparable (500 mg L^{-1}) doses of application, Eck [19] and Howell et al. [52] reported that ethephon did not affect TSS in highbush blueberry, consistent with results from the current study. In rabbiteye blueberry, Dekazos [20] reported no effect of ethephon on TSS even with repeated 500 mg L^{-1} applications or a single 1000 mg L^{-1} application. However, Ban et al. [21] reported an increase in TSS with ethephon applications at 8 days after treatment in Tifblue, although the effects of this application on postharvest storage were not evaluated in this study. Overall, it appears that ethephon applications do not generally alter TSS content in blueberry fruit during postharvest storage. Several studies have reported a decrease in TA after ethephon applications (ranging from 200 to 3840 mg L^{-1}) in highbush and rabbiteye blueberry [19–21]. In the current study, TA levels were unaffected by ethephon in Premier and slightly increased in Powderblue during postharvest storage (15 days). It is possible that the genotypes used here responded differently for this ripening parameter. Juice pH was generally not affected by ethephon treatment as has been seen previously [19,20], except at 1 day after storage. Dekazos [20] reported changes in fruit color parameters in response to ethephon, in contrast to that reported here. As indicated above, Dekazos [20] used repeated and higher doses of ethephon, which may explain the different results observed. Overall, data from this study suggest that ethephon application at 250 mg L^{-1} may have minimal effects on rabbiteye blueberry fruit quality during postharvest storage.

The role of ethylene in regulating blueberry ripening and postharvest quality is not completely clear [50]. Although a peak in respiration and ethylene production has been observed in blueberry in some studies [6,17,18], this was not conclusive in others [22]. Treatment of mature fruit with the ethylene perception inhibitor 1-MCP enhanced ethylene production, accelerated loss of fruit firmness, and had little effect on fruit quality characteristics in rabbiteye blueberry cultivars [50], underlining the complex and unclear role of ethylene in regulating blueberry ripening. Recently, based on the analysis of transcriptomics data during various stages of fruit development in highbush blueberry, Gupta et al. [53] indicated that genes associated with ethylene biosynthesis were abundant during the initiation of ripening, suggesting that ethylene may in fact play specific role(s) in modulating the progression of this process in blueberry fruit. The data presented in the current study demonstrating the effect of an ethylene-releasing compound on the progression of ripening further support a potential role for ethylene in the regulation of the ripening program in rabbiteye blueberry. These data indicate that blueberry fruit are responsive to external ethylene. Further studies evaluating the climacteric/non-climacteric nature of blueberry ripening are needed to better understand the potential role of ethylene in the regulation of this fruit developmental process. This information will also be critical for fine-tuning the timing of ethephon application in relation to fruit development.

In the current study, ABA generally did not affect the rate of ripening in blueberry fruit, even when applied at a rate of 1000 mg L^{-1}, although it increased the proportion of pink fruit. It may be likely that ABA (1000 mg L^{-1}) was able to stimulate the synthesis of anthocyanin pigments associated with pink color in the fruit. Phytotoxicity symptoms were observed in leaves when ABA was applied at 1000 mg L^{-1} (data not shown). Application of ABA did not consistently affect any of the fruit quality characteristics measured across the two cultivars except for compression in Premier at 29 days and TSS in Powderblue at 15 days after harvest. Furthermore, ABA applications appeared to increase the proportion of defective fruit at 29 days after storage in Premier. Overall, external ABA applications did not influence the progression of ripening in rabbiteye blueberry, in contrast to some previous reports with highbush blueberry and closely-related bilberry, where ABA concentration was found to increase during ripening leading to the hypothesis that it may regulate anthocyanin biosynthesis and other ripening related characteristics [35,36]. It may be that rabbiteye blueberry is less responsive to ABA or that the genotypes studied exhibited limited ABA responsiveness. Further analysis involving comparison of different blueberry genotypes and species may be needed to clarify any potential roles of ABA in blueberry ripening.

While MeJA applications have been noted to alter the progression of ripening in several fruits such as strawberry, raspberry, peach, apple, and tomato [14,39,43], in the current study MeJA application did not result in any consistent effects on the progression of ripening even when the application doses were at 1 mM. Further evaluation may be required to determine whether higher doses of MeJA can affect ripening in blueberry. However, previous studies have indicated that MeJA application at 10 mM and higher accelerate fruit detachment and result in extensive fruit drop in blueberry [54–56]. Hence, if higher doses of MeJA are successful at accelerating the progression of ripening, this needs to be optimized such that fruit detachment responses are not induced. Additionally, quantification of jasmonates during fruit development and specifically during ripening may help provide further insights into their potential roles in blueberry ripening.

5. Conclusions

Data from this study indicated that ethephon applied at a relatively low dose of 250 mg L^{-1} accelerated the progression of ripening in rabbiteye blueberries without altering many of the fruit quality characteristics. The other two PGRs tested, ABA and MeJA, did not appear to alter the progression of ripening in these cultivars. Further studies are needed to determine whether ethephon can consistently alter the ripening process in other types of blueberry, particularly southern highbush blueberry. In such studies, it may be essential to evaluate multiple doses of application, stages of application in relation to fruit development, and the time of day of application to determine the optimum application parameters for this PGR. Additionally, considering the response to ethephon, mechanisms involved in ethylene-mediated alteration of fruit ripening warrant further evaluation in blueberry.

References

1. Norberto, S.; Silva, S.; Meireles, M.; Faria, A.; Pintado, M.; Calhau, C. Blueberry anthocyanins in health promotion: A. metabolic overview. *J. Funct. Foods* **2013**, *5*, 1518–1528. [CrossRef]

2. Evans, E.A.; Ballen, F.H. *An Overview of US Blueberry Production, Trade, and Consumption, with Special Reference to Florida*; Publication FE952; University of Florida, Institute of Food and Agricultural Sciences: Gainesville, FL, USA, 2014.

3. Lang, G.A. Southern highbush blueberries: Physiological and cultural factors important for optimal cropping of these complex hybrids. *Acta Hortic.* **1993**, *346*, 72–80. [CrossRef]

4. Boches, P.S.; Bassil, N.V.; Rowland, L.J. Genetic diversity in the highbush blueberry evaluated with microsatellite markers. *J. Am. Soc. Hort. Sci.* **2006**, *131*, 674–686.

5. Rowland, L.J.; Ogden, E.L.; Bassil, N.; Buck, E.J.; McCallum, S.; Graham, J.; Brown, A.; Wiedow, C.; Campbell, A.M.; Haynes, K.G.; et al. Construction of a genetic linkage map of an interspecific diploid blueberry population and identification of QTL for chilling requirement and cold hardiness. *Mol. Breed.* **2014**, *34*, 2033–2048. [CrossRef]

6. Suzuki, A.; Kikuchi, T.; Aoba, K. Changes of ethylene activity evolution, ACC content, in fruits ethylene forming blueberry enzyme and respiration of highbush blueberry. *J. Jpn Soc. Hort. Sci.* **1997**, *66*, 23–27. [CrossRef]

7. Brown, G.K.; Marshall, D.E.; Tennes, B.R.; Booster, D.E.; Chen, P.; Garrett, R.E.; O'Brien, M.; Studer, H.E.; Kepner, R.A.; Hedden, S.L.; et al. Status of harvest mechanization of horticultural crops. *Am. Soc. Agric. Eng.* **1983**, *83*, 3.

8. Brown, G.K.; Shulte, N.L.; Timm, E.J.; Beaudry, R.M.; Peterson, D.L.; Hancock, J.F.; Takeda, F. Estimates of mechanization effects on fresh blueberry quality. *Appl. Eng. Agric.* **1996**, *12*, 21–26. [CrossRef]

9. Safley, C.D.; Cline, W.O.; Mainland, C.M. Estimated costs of producing, harvesting, and marketing blueberries in the southeastern United States. In Proceedings of the 12th Biennial Southeast Blueberry Conference, Savannah, GA, USA, 6–9 January 2005; pp. 33–49.

10. McAtee, P.; Karim, S.; Schaffer, R.; David, K. A dynamic interplay between phytohormones is required for fruit development, maturation, and ripening. *Front. Plant Sci.* **2013**, *4*, 79. [CrossRef] [PubMed]

11. Giovannoni, J.J. Genetic regulation of fruit development and ripening. *Plant Cell* **2004**, *16*, 170–181. [CrossRef] [PubMed]

12. Symons, G.M.; Chua, Y.J.; Ross, J.J.; Quittenden, L.J.; Davies, N.W.; Reid, J.B. Hormonal changes during non-climacteric ripening in strawberry. *J. Expt. Bot.* **2012**, *63*, 4741–4750. [CrossRef] [PubMed]

13. Abdul Shukor, A.R.; Yulianingsih, N.H.; Acedo, A.L.; Teng, K.C. Regulation of ripening in banana. In *Banana: Fruit Development, Postharvest Physiology, Handling and Marketing in ASEAN*; Abdullah, H., Pantastico, E.B., Eds.; ASEAN Food Handling Bureau: Kuala Lumpur, Malaysia, 1990; pp. 72–84.

14. Cherian, S.; Figueroa, C.R.; Nair, H. 'Movers and shakers' in the regulation of fruit ripening: A cross-dissection of climacteric versus non-climacteric fruit. *J. Expt. Bot.* **2014**, *65*, 4705–4722. [CrossRef] [PubMed]

15. Giovannoni, J. Molecular biology of fruit maturation and ripening. *Annu. Rev. Plant Physiol. Plant Mol. Biol.* **2001**, *52*, 725–749. [CrossRef] [PubMed]

16. Klee, H.J.; Giovannoni, J.J. Genetics and control of tomato fruit ripening and quality attributes. *Annu. Rev. Genet.* **2011**, *45*, 41–59. [CrossRef] [PubMed]

17. Windus, N.D.; Shutak, V.G.; Gough, R.E. CO_2 and C_2H_4 evolution by highbush blueberry fruit. *HortScience* **1976**, *11*, 515–517.

18. El-Agamy, S.; Aly, M.M.; Biggs, R.H. Fruit maturity as related to ethylene in 'Delite' blueberry. *Proc. Fla. State Hortic. Soc.* **1982**, *95*, 245–246.

19. Eck, P. Influence of ethrel upon highbush blueberry fruit ripening. *HortScience* **1970**, *5*, 23–25.

20. Dekazos, E.D. Effects of preharvest applications of ethephon and SADH on ripening, firmness and storage quality of rabbiteye blueberries (cv 'T.-19'). *Proc. Fla. State Hort. Soc.* **1976**, *89*, 266–270.

21. Ban, T.; Kugishima, M.; Ogata, T.; Shiozaki, S.; Horiuchi, S.; Ueda, H. Effect of ethephon (2-chloroethylphosphonic acid) on the fruit ripening characters of rabbiteye blueberry. *Sci. Hortic.* **2007**, *112*, 278–281. [CrossRef]

22. Frenkel, C. Involvement of peroxidase and indole-3-acetic acid oxidase isozymes from pear, tomato, and blueberry fruit in ripening. *Plant Physiol.* **1972**, *49*, 757–763. [CrossRef] [PubMed]

23. Buesa, C.; Dominguez, M.; Vendrell, M. Abscisic acid effects on ethylene production and respiration rate in detached apple fruits at different stages of development. *Revista Española de Ciencia y Tecnología de Alimentos* **1994**, *34*, 495–506.

24. Rodrigo, M.; Marcos, J.F.; Alferez, F.; Mallent, M.D.; Zacarias, L. Characterization of Pinalate, a novel *Citrus sinensis* mutant with a fruit-specific alteration that results in yellow pigmentation and decreased ABA content. *J. Exp. Bot.* **2003**, *54*, 727–738. [CrossRef] [PubMed]

25. Kondo, S.; Inoue, K. Abscisic acid (ABA) and 1-aminocyclopropane-1-carboxylic acid (ACC) content during growth of 'Satohnishiki' cherry fruit, and effect of ABA and ethephon application on fruit quality. *J. Hortic. Sci.* **1997**, *72*, 221–227. [CrossRef]

26. Chai, Y.M.; Jia, H.F.; Li, C.L.; Dong, Q.H.; Shen, Y.Y. FaPYR1 is involved in strawberry fruit ripening. *J. Expt. Bot.* **2011**, *62*, 5079–5089. [CrossRef] [PubMed]

27. Jia, H.F.; Chai, Y.M.; Li, C.L.; Lu, D.; Luo, J.J.; Qin, L.; Shen, Y.Y. Abscisic acid plays an important role in the regulation of strawberry fruit ripening. *Plant Physiol.* **2011**, *157*, 188–199. [CrossRef] [PubMed]

28. Gagné, S.; Esteve, K.; Deytieux-Belleau, C.; Saucier, C.; Geny, L. Influence of abscisic acid in triggering véraison in grape berry skins of *Vitis vinifera* L. cv. Cabernet-Sauvignon. *J. Int. Sci.* **2006**, *40*, 7–14. [CrossRef]

29. Jeong, S.; Goto-Yamamoto, N.; Kobayashi, S.; Esaka, M. Effects of plant hormones and shading on the accumulation of anthocyanins and the expression of anthocyanin biosynthetic genes in grape berry skins. *Plant Sci.* **2004**, *167*, 247–252. [CrossRef]

30. Peppi, M.C.; Fidelibus, M.W.; Dokoozlian, N. Abscisic acid application timing and concentration affect firmness, pigmentation, and color of 'Flame Seedless' grapes. *HortScience* **2006**, *41*, 1440–1445.

31. Cantín, C.M.; Fidelibus, M.W.; Crisosto, C.H. Application of abscisic acid (ABA) at Veraison advanced red color development and maintained postharvest quality of "Crimson Seedless" grapes. *Postharvest Biol. Technol.* **2007**, *46*, 237–241. [CrossRef]

32. Zhang, M.; Yuan, B.; Leng, P. The role of ABA in triggering ethylene biosynthesis and ripening of tomato fruit. *J. Exp. Bot.* **2009**, *60*, 1579–1588. [CrossRef] [PubMed]

33. Mou, W.; Li, D.; Bu, J.; Jiang, Y.; Khan, Z.U.; Luo, Z.; Mao, L.; Ying, T. Comprehensive analysis of ABA effects on ethylene biosynthesis and signaling during tomato fruit ripening. *PLoS ONE* **2016**, *11*, e0154072. [CrossRef] [PubMed]

34. Jiang, Y.; Joyce, D.; Macnish, A. Effect of abscisic acid on banana fruit ripening in relation to the role of ethylene. *J. Plant Growth Regul.* **2000**, *19*, 106–111. [CrossRef] [PubMed]

35. Karppinen, K.; Hirvelä, E.; Nevala, T.; Sipari, N.; Suokas, M.; Jaakola, L. Changes in the abscisic acid levels and related gene expression during fruit development and ripening in bilberry (*Vaccinium myrtillus* L.). *Phytochemistry* **2013**, *95*, 127–134. [CrossRef] [PubMed]

36. Zifkin, M.; Jin, A.; Ozga, J.A.; Zaharia, L.I.; Schernthaner, J.P.; Gesell, A.; Abrams, S.R.; Kennedy, J.A.; Constabel, C.P. Gene expression and metabolite profiling of developing Highbush blueberry fruit indicates transcriptional regulation of flavonoid metabolism and activation of abscisic acid metabolism. *Plant Physiol.* **2012**, *158*, 200–224. [CrossRef] [PubMed]

37. Buran, T.J.; Sandhu, A.K.; Azeredo, A.M.; Bent, A.H.; Williamson, J.G.; Gu, L. Effects of exogenous abscisic acid on fruit quality, antioxidant capacities, and phytochemical contents of southern high bush blueberries. *Food Chem.* **2012**, *132*, 1375–1381. [CrossRef] [PubMed]

38. Wasternack, C.; Hause, B. Jasmonates: Biosynthesis, perception, signal transduction and action in plant stress response, growth and development. *Ann. Bot.* **2013**, *111*, 1021–1058. [CrossRef] [PubMed]

39. Rohwer, C.L.; Erwin, J.E. Horticultural applications of jasmonates: A review. *J. Hortic. Sci. Biotechnol.* **2008**, *83*, 283–304. [CrossRef]

40. Fan, X.; Mattheis, J.P.; Fellman, J.K. A role for jasmonates in climacteric fruit ripening. *Planta* **1998**, *204*, 444–449. [CrossRef]

41. Kondo, S.; Setha, S.; Rudell, D.R.; Buchanan, D.; Mattheis, J.P. Aroma volatile biosynthesis in apples affected by 1-MCP and methyl jasmonate. *Postharvest Biol. Technol.* **2005**, *36*, 61–68. [CrossRef]

42. Ziosi, V.; Bonghi, C.; Bregoli, A.M.; Trainotti, L.; Biondi, S.; Sutthiwal, S.; Kondo, S.; Costa, G.; Torrigiani, P. Jasmonate-induced transcriptional changes suggest a negative interference with the ripening syndrome in peach fruit. *J. Exp. Bot.* **2008**, *59*, 563–573. [CrossRef] [PubMed]

43. Wei, J.; Wen, X.; Tang, L. Effect of methyl jasmonic acid on peach fruit ripening progress. *Sci. Hortic.* **2017**, *220*, 206–213. [CrossRef]

44. Pérez, A.G.; Sanz, C.; Olías, R.; Olías, J.M. Effect of methyl jasmonate on in vitro strawberry ripening. *J. Agric. Food Chem.* **1997**, *45*, 3733–3737. [CrossRef]

45. Concha, C.M.; Figueroa, N.E.; Poblete, L.A.; Oñate, F.A.; Schwab, W.; Figueroa, C.R. Methyl jasmonate treatment induces changes in fruit ripening by modifying the expression of several ripening genes in *Fragaria chiloensis* fruit. *Plant Physiol. Biochem.* **2013**, *70*, 433–444. [CrossRef] [PubMed]

46. Wang, S.Y.; Zheng, W. Preharvest application of methyl jasmonate increases fruit quality and antioxidant capacity in raspberries. *Int. J. Food Sci. Technol.* **2005**, *40*, 187–195. [CrossRef]

47. Cocetta, G.; Rossoni, M.; Gardana, C.; Mignani, I.; Ferrante, A.; Spinardi, A. Methyl jasmonate affects phenolic metabolism and gene expression in blueberry (*Vaccinium corymbosum*). *Physiol. Plant.* **2015**, *153*, 269–283. [CrossRef] [PubMed]

48. Mehra, L.K.; MacLean, D.D.; Savelle, A.T.; Scherm, H. Postharvest disease development on southern highbush blueberry fruit in relation to berry flesh type and harvest method. *Plant Dis.* **2013**, *97*, 213–221. [CrossRef]

49. Forsyth, F.R.; Craig, D.L.; Stark, R. Ethephon as a chemical harvesting aid for the late season highbush blueberry Coville. *Can. J. Plant Sci.* **1977**, *57*, 1099–1102. [CrossRef]

50. MacLean, D.D.; NeSmith, D.S. Rabbiteye blueberry postharvest fruit quality and stimulation of ethylene production by 1-Methylcylopropene. *HortScience* **2011**, *46*, 1278–1281.

51. Concha-Meyer, A.; Eifert, J.D.; Williams, R.C.; Marcy, J.E.; Welbaum, G.E. Shelf life determination of fresh blueberries (*Vaccinium corymbosum*) stored under controlled atmosphere and ozone. *Int. J. Food Sci.* **2015**. [CrossRef] [PubMed]

52. Howell, G.S.; Stergios, B.G.; Stackhouse, S.S.; Bittenbender, H.C.; Burton, C.L. Ethephon as a mechanical harvesting aid for highbush blueberries (*Vaccinium austral* Small). *J. Am. Soc. Hort. Sci.* **1976**, *101*, 111–115.

53. Gupta, V.; Estrada, A.D.; Blakley, I.; Reid, R.; Patel, K.; Meyer, M.D.; Andersen, S.U.; Brown, A.F.; Lila, M.A.; Loraine, A.E. RNA-Seq analysis and annotation of a draft blueberry genome assembly identifies candidate genes involved in fruit ripening, biosynthesis of bioactive compounds, and stage-specific alternative splicing. *GigaScience* **2015**, *4*, 5. [CrossRef] [PubMed]

54. Malladi, A.; Vashisth, T.; Johnson, L.K. Ethephon and methyl jasmonate affect fruit detachment in rabbiteye and southern highbush blueberry. *HortScience* **2012**, *47*, 1745–1749.

55. Vashisth, T.; Malladi, A. Fruit detachment in rabbiteye blueberry: Abscission and physical separation. *J. Am. Soc. Hortic. Sci.* **2013**, *138*, 95–101.

56. Vashisth, T.; Malladi, A. Fruit abscission in rabbiteye blueberry in response to organ removal and mechanical wounding. *HortScience* **2014**, *49*, 1403–1407.

Chapter-5

Effect of Continuous Exposure to Low Levels of Ethylene on Mycelial Growth of Postharvest Fruit Fungal Pathogens

1. Introduction

Ethylene is a gaseous plant hormone that is well known to initiate postharvest ripening of climacteric fruit, and enhance the senescence of non-climacteric fruit and vegetables [1]. A common feature of such postharvest changes in fruits and vegetables is an increased susceptibility to develop fungal decay [2,3]. Published data on the effect of ethylene on fungal decay development have mostly been in vivo studies, where native or inoculated produce were exposed to atmospheric ethylene at a wide range of concentrations. Early interest in the role of ethylene arose from the commercial postharvest degreening process of citrus fruit where physiologically mature but green skinned fruit are exposed to relatively high temperatures (25 to 30 °C) in the presence of an ethylene concentration of about 5 μL L^{-1} to accelerate the loss of chlorophyll in the peel and thus 'degreen' the fruit [4,5]. The findings on the effect of such ethylene concentrations on postharvest decay of citrus fruit are inconsistent. Grierson and Newhall [6] found enhanced incidence of stem-end rot due to *Diplodia natalensis*, and Brown [7] found reduced green mould due to *Penicillium digitatum*; while Plaza et al. [8] found no effect on mould incidence for *P. digitatum*, and Porat et al. [9] showed decreased blue mould due to *P. italicum*. The effect of ethylene on the development of decay in other fruits include Zhu et al. [10] who reported that fumigation of 200 μL L^{-1} ethylene had no effect on grey mould incidence in grapes inoculated with *Botrytis cinerea*, Palou et al. [11] who found that decay development from *Monilinia fructicola* inoculated into a range of stone fruits was not affected by exposure to ethylene up to 100 μL L^{-1}, and Lockhart et al. [12] who reported that exposure to ethylene at about 2000 μL L^{-1} inhibited development of apple rot caused by *Gloeosporium album*. Thus in vivo studies have shown variable effects of ethylene on fungal rot development of fruits. However, it is not possible from these studies to ascertain whether any enhanced mould growth induced by ethylene is a secondary effect arising from increased cell permeability that facilitates germination and growth of fungal spores [13], or any inhibition of mould growth is due to a defence response by tissues to ethylene [14].

An alternate action, that needs to be considered, is whether ethylene directly affects the growth of fungi present on the surface of produce as speculated by Sharon et al. [15] for *Botrytis cinerea*. It would seem necessary to conduct in vitro studies in order to determine if ethylene has a direct effect on fungal growth. However, few such studies have been reported. Lockhart et al. [12] reported in vitro growth of *Gloeosporium album* was not affected by exposure to 2000 μL L^{-1} ethylene while Kepczynski and Kepczynska [16] reported that germination of *B. cinerea*, *P. expansum*, *Gloeosporium perennans* and *Rhizopus nigricans* was stimulated by exposure to ethylene at 10 to 1000 μL L^{-1}. El-Kazzaz et al. [17] exposed 10 pathogens to ethylene at 1 to 1000 μL L^{-1} and found fungal growth rates were not significantly affected by ethylene for *Alternaria alternata*, *Botryodiplodia theobromae*, *B. cinerea*, *P. digitatum*, and *P. expansum* while ethylene showed a slight but significant increase between some ethylene concentrations for *Colletotrichum gloeosporioides*, *M. fructicola*, *P. italicum*, *Rhizopus stolonifera*, and *Thielaviopsis paradoxa*. None of these studies indicated the fruit source of the fungi. The only study to specify the fungal source was Brown and Lee [13] who found growth of *D. natalensis* isolated from Temple orange was inhibited by exposure to 55 μL L^{-1} ethylene.

Most of the in vitro and in vivo studies have been conducted by exposing produce to a much higher concentration of ethylene than would normally be encountered during commercial marketing situations which has been found to be invariably <2 μL L^{-1} and commonly about 0.1 μL L^{-1} [18,19]. While ethylene concentrations may be higher in internal produce tissue, many fungal pathogens reside on the surface of the produce and are therefore exposed to atmospheric concentrations of ethylene. Any action of abnormally high ethylene concentrations of ethylene (>2 μL L^{-1}) is of scientific interest but the response to lower ethylene concentrations would better demonstrate how fungal pathogens behave on fruit and vegetables in commercial supply chains. In this study, fungal spores were isolated from eight ripe fruits. Actively growing mycelia taken from rots on these fruits were sub-cultured onto a petri dish and continually ventilated at 20 °C with an air stream containing 0 (actually <0.001 μL L^{-1}), 0.1 and 1 μL L^{-1} ethylene. The growth of fungal mycelia was observed daily and measured when the mycelial diameter for each pathogen was about 50% of the petri dish diameter.

2. Materials and Methods

Persimmon, pear, tomato, mango, papaya, orange, grape, and blueberry fruit of commercial maturity with a visible rot were obtained from a retail outlet. The pathogen was collected from the rot area by cutting a disc (4 mm diam.) of flesh under the fruit skin at the outer edge of the lesion. The pathogen disc from each fruit was cultured on potato dextrose agar (PDA) (Difco Laboratories, Sparks, MD, USA). PDA media (120 mL) was placed into each petri dish (90 mm diam.) using a liquid dispenser (OminispensePlus, Millville, NJ, USA). A petri dish was inoculated in the centre with the pathogen disc and incubated at 25° C for 3 to 5 days to allow pathogen development. A sub-culture was then obtained from the edge of the mycelia with a core borer (6 mm diam.). The sub-sample for each pathogen was re-plated onto three petri dishes and incubated at 25 °C for two days to grow in ambient air before the petri dishes were transferred into ethylene treatments. However, the pathogen isolate obtained from tomatoes were transferred immediately to ethylene storage system without the pre-incubation process as the pathogen grew more rapidly than the other isolates. Groups of three petri dishes with the same fruit isolate constituted a replicate and were placed into three separate metal drums (60 L) that were fitted with inlet and outlet ports in the lid. Containers were placed into a temperature controlled room at 20 °C. Each fruit pathogen was assessed with four to nine independent replicate trials at different times.

Containers were ventilated with humidified air (100 mL min^{-1}) containing the concentrations of ethylene of 0, 0.1 or 1 μL L^{-1}. The desired ethylene level was obtained by mixing ethylene from a gas cylinder (BOC Gases, Sydney, Australia) with compressed air that was made "ethylene-free" by passing through a jar containing potassium permanganate pellets, and humidified to approximately 95% RH by bubbling through water. Ethylene concentration flowing through the system was monitored daily by flame ionisation gas chromatography as described by Huque et al. [20] and to be the same inside the petri dish as in the surrounding atmosphere. The "ethylene-free" air contained <0.001 μL L^{-1} ethylene as this was the limit of detection of the analytical method. Growth of fungal mycelia was observed daily and measured when the mycelial diameter for each pathogen was in the range 30–50 mm which was about 50% of the petri dish diameter. On the designated assessment day, petri dishes were removed from the ethylene system and placed under sterile conditions where the diameter of mycelia was measured twice diagonally, and an average diameter calculated for each plate. The data for mycelial growth of each pathogen were subjected to analysis of variance using the statistical analysis system (SAS) 9.4 program. When the mean for growth of a fungal organism was found to be significantly different between ethylene levels, the least significant difference (LSD) at $P = 0.05$ was calculated.

Fungi associated with diseased lesions on each of the eight ripe fruit were isolated to agar media and identified to genus or species using conventional morphological taxonomy. In all cases a single fungus free of contamination was observed. Their identities were then confirmed or further characterised by comparison of their ITS rDNA sequences with GenBank accessions. Cultures were also submitted to the NSW plant pathogen herbarium (DAR). All identified fungi are known pathogens on the respective host fruit and are listed in Table 1.

Table 1. Mycelial growth of postharvest fungi isolated from eight fruit in the presence of low concentrations of ethylene.

Pathogen	Fruit	Cultivar/Type	Reps *	Days †	Mycelial Growth (mm diam) 0	0.1	1 µL L−1 C2H4	P Value	LSD ‡
Geotrichum candidum	Persimmon	Climacteric Fruit Fuyu	8	13	50.3 a	45.9 b	45.9 b	0.003	2.5
Cladosporium herbarum	Pear	Packham	4	13	45.8 a	39.9 b	40.9 b	0.000	3.0
Rhizopus stolonifera	Tomato	Gourmet	5	2	67.9 a	63.8 b	63.6 b	0.041	3.7
Colletotrichum asianum	Mango	Kensington Pride	6	5	39.3 a	35.8 b	36.2 b	0.031	2.8
Phomopsis caricae-papayae	Papaya	Yellow papaw	6	13	49.1 a	44.6 b	43.0 b	0.004	3.5
		Non-Climacteric Fruit							
Penicillium waksmanii	Orange	Washington Navel	9	9	40.3	41.4	40.0	0.84	ns
Botrytis cinerea	Grape	Black seedless	8	3	32.5	30.4	30.6	0.09	ns
Colletotrichum simmondsii	Blueberry	OB 1	8	9	37.4	36.0	35.6	0.21	ns

** Number of replicated trials using the same batch of fungal spores. † Number of days at 20°C before assessment. ‡ Least significant difference between means at P = 0.05.*

3. Results and Discussion

Table 1 gives the time for each pathogen to grow to 30–50 mm and the effect of exposure to atmospheric ethylene at 0, 0.1 and 1 μL L^{-1} on in vitro mycelial growth at this time. The results show there was a significant effect of ethylene on the growth of five pathogens, *Geotrichum candidum*, *Cladosporium herbarum*, *Rhizopus stolonifera*, *Colletotrichum asianum*, and *Phomopsis caricae-papayae*. Mycelial growth of these fungi was greatest when they were in an ethylene-free (i.e., <0.001 μL L^{-1}) atmosphere. Pathogen growth was significantly decreased when held in an ethylene concentration of 0.1 μL L^{-1} but there was no further decrease in growth in the presence of 1 μL L^{-1} ethylene. The other three pathogens, *Penicillium waksmanii*, *Botrytis cinerea*, and *Colletotrichum simmondsii*, did not show any significant difference in mycelial growth in the ethylene concentrations.

The five pathogens that showed inhibited growth in the presence of exogenous ethylene were, respectively, isolated from persimmon, pear, tomato, mango and papaya (Table 1), which are all climacteric fruits. Climacteric fruit are defined by having a distinct increase in endogenous ethylene and respiration associated with ripening [1]. The three non-ethylene responsive pathogens were, respectively, isolated from orange, grape and blueberry (Table 1), which are all non-climacteric fruits and do not exhibit a climacteric rise in ethylene and respiration after harvest. This result was quite unexpected. It would seem logical that ethylene-tolerant pathogens would preferentially colonise climacteric fruits. The inhibitory effect of ethylene on growth of climacteric fruit pathogens raises an evolutionary question of what advantage does the organism derive from having its growth rate slightly inhibited when the fruit starts producing ethylene and hence begins ripening. A possible explanation needs to be viewed in the context that endogenous ethylene stimulates loss of cellular integrity of the fruit as part of the ripening process, and the loss of cellular integrity facilitates access by mycelia to nutrients in fruit tissues. A slight delay in fungal growth could be advantageous to allow greater degradation of fruit tissues to occur so that colonisation by the pathogen can occur more rapidly thereafter. The lack of any response by fungal pathogens that colonise non-climacteric fruit to an increase in ethylene production could reflect the lack of a need to adapt to a natural increase in ethylene production by such fruits early in the senescence process.

It would seem that exposure of climacteric fruit fungal pathogens to an ethylene concentration of 0.1 μL L^{-1} generates the greatest inhibition of fungal growth as no pathogen showed any further growth inhibition when exposed to 1.0 μL L^{-1} ethylene. An ethylene concentration of 0.1 μL L^{-1} was found by Wills et al. [18] to be a common background concentration of fruit and vegetables in a wholesale market. However, the magnitude of the reduction in mycelial growth of the climacteric fruit pathogens in the presence of 0.1 and 1.0 μL L^{-1} ethylene was only about 10% which would not be great enough to be of commercial significance. The longest postharvest life would still be attained by maintaining ethylene as low as possible as this will delay the loss of cellular integrity that enhances to the ability of fungi to grow.

It is recognised that only a limited number of fruits were examined in this report and an extended study on a greater range of fruits with exposure to a greater range of ethylene concentrations is warranted to test the universality of the findings. It is further recognised that some of the fungi tested in the study have a very broad host range, for example, *Botrytis cinerea* and *Geotrichum candidum* can invade both climacteric and non-climacteric fruit. Further research could examine the ethylene response to isolates of broad host fungi from different produce to see if there is a differential adaptive response to ethylene. Such studies could also examine if the ethylene response to pathogen mycelial growth also corresponds with infection potential by measuring parameters such as spore germination, appressorial formation and germ tube elongation or even to the ability of different fungi to metabolise ethylene.

4. Conclusions

The finding that pathogens isolated from all the climacteric fruits showed reduced growth when exposed to a low level of ethylene was quite unexpected. It suggests that pathogens colonising climacteric fruits derive some advantage by delaying growth early in the ripening sequence, possibly related to enhanced loss of cellular integrity arising from increasing ethylene action on fruit tissues. The lack of an effect of ethylene on pathogens from non-climacteric fruit could reflect the lack of enhanced ethylene synthesis during the ripening of such fruits and thus there is no advantage in an adaptive response to ethylene.

References

1. Wills, R.B.H.; Golding, J.B. *Postharvest: An Introduction to the Physiology and Handling of Fruit and Vegetables*, 6th ed.; UNSW Press: Sydney, Australia, 2016.
2. Kader, A.A. Ethylene-induced senescence and physiological disorders in harvested horticultural crops. *HortScience* **1985**, *20*, 54–57.
3. Martínez-Romero, D.; Bailén, G.; Serrano, M.; Guillén, F.; Valverde, J.M.; Zapata, P.; Castillo, S.; Valero, D. Tools to maintain postharvest fruit and vegetable quality through the inhibition of ethylene action: A review. *Crit. Rev. Food Sci. Nutr.* **2007**, *47*, 543–560. [CrossRef] [PubMed]
4. Wardowski, W.F.; McCornack, A.A. *Recommendations for Degreening Florida Fresh Citrus Fruits*; Florida Cooperative Extension Service: Gainsville, FL, USA, 1973.
5. Ritenour, M.A.; Miller, W.M.; Wardowski, W.W. Recommendations for Degreening Florida Fresh Citrus Fruits. In *Institute of Food and Agricultural Sciences Extension Circular 1170*; Horticultural Sciences Department, University of Florida: Gainesville, FL, USA, 2003.
6. Grierson, W.; Newhall, W.F. Tolerance to ethylene of various types of citrus fruit. *Proc. Am. Soc. Hortic. Sci.* **1955**, *65*, 244–250.
7. Brown, G.E. Development of green mold in degreened oranges. *Phytopathology* **1973**, *63*, 1104–1107. [CrossRef]
8. Plaza, P.; Sanbruno, A.; Usall, J.; Lamarca, N.; Torres, R.; Pons, J.; Viñas, I. Integration of curing treatments with degreening to control the main postharvest diseases of clementine mandarins. *Postharvest Biol. Technol.* **2004**, *34*, 29–37. [CrossRef]

9. Porat, R.; Weiss, B.; Cohen, L.; Daus, A.; Goren, R.; Droby, S. Effects of ethylene and 1-methylcyclopropene on the postharvest qualities of 'Shamouti' oranges. *Postharvest Biol. Technol.* **1999**, *15*, 155–163. [CrossRef]
10. Zhu, P.K.; Xu, L.; Zhang, C.H.; Toyoda, H.; Gan, S.S. Ethylene produced by botrytis cinerea can affect early fungal development and can be used as a marker for infection during storage of grapes. *Postharvest Biol. Technol.* **2012**, *66*, 23–29. [CrossRef]
11. Palou, L.X.; Crisosto, C.H.; Garner, D.; Basinal, L.M. Effect of continuous exposure to exogenous ethylene during cold storage on postharvest decay development and quality attributes of stone fruits and table grapes. *Postharvest Biol. Technol.* **2003**, *27*, 243–254. [CrossRef]
12. Lockhart, C.L.; Forsyth, F.R.; Eaves, C.A. Effect of ethylene on development of gloeosporium album in apple and on growth of the fungus culture. *Can. J. Plant Sci.* **1968**, *48*, 557–559.
13. Brown, G.E.; Lee, H.S. Interaction of ethylene with citrus stem-end rot caused by *Diplodia natalensis*. *Phytopathology* **1993**, *83*, 1204–1208. [CrossRef]
14. Saltveit, M.E. Effect of ethylene on quality of fresh fruits and vegetables. *Postharvest Biol. Technol.* **1999**, *15*, 279–292. [CrossRef]
15. Sharon, A.; Elad, Y.; Barakat, R.; Tudzynski, P. Phytohormones in Botrytis–Plant Interactions. In *Botrytis: Biology, Pathology and Control*; Elad, Y., Williamson, B., Tudzynski, P., Delen, N., Eds.; Springer: Dordrecht, The Netherlands, 2004; pp. 163–179.
16. Kepczynski, J.; Kepczynska, E. Effect of ethylene on germination of fungal spores causing fruit rot. *Fruit Sci. Rep.* **1977**, *4*, 31–35.
17. El-Kazzaz, M.K.; Sommer, N.F.; Kader, A.A. Ethylene effects on in vitro and in vivo growth of certain postharvest fruit-infecting fungi. *Phytopathology* **1983**, *73*, 998–1001. [CrossRef]
18. Wills, R.B.H.; Warton, M.A.; Ku, V.V.V. Ethylene levels associated with fruit and vegetables during marketing. *Aust. J. Exp. Agric.* **2000**, *40*, 465–470.
19. Li, Y.; Wills, R.B.H.; Golding, J.B. Interaction of ethylene concentration and storage temperature on postharvest life of the green vegetables pak choi, broccoli, mint, and green bean. *J. Hortic. Sci. Biotechnol.* **2017**, *92*, 288–293. [CrossRef]
20. Huque, R.; Wills, R.B.H.; Pristijono, P.; Golding, J.B. Effect of nitric oxide (no) and associated control treatments on the metabolism of fresh-cut apple slices in relation to development of surface browning. *Postharvest Biol. Technol.* **2013**, *78*, 16–23. [CrossRef]

Chapter-6

The Potential Use of Hot Water Rinsing and Brushing Technology to Extend Storability and Shelf Life of Sweet Acorn Squash (*Cucurbita pepo* L.)

1. Introduction

Acorn squash are an old cultivar-group of *Cucurbita pepo* L. subsp. *texana* (Scheele) Filov [1]. The fruits have a distinctive turbinate shape with several longitudinal ridges and furrows [2], and they are rather small, weighing 300–1500 g. Most acorn squash cultivars have a dark green rind with a deep-orange flesh and sugar content of 12%–18% [3]. The fruit is an excellent source of nutrients and vitamins and can be stored for about 2 months at temperatures not lower than 10 °C without developing chilling injury and losing its dark color, which is a good indicator for freshness and marketing [4]. Under the hot Israeli climate, the fruits lose their dark-green rind color, become dull orange or yellow within a short time after harvest, and thereby lose value. Because of its dark green color and distinctive taste after cooking, this fruit is called Dela'at Armonim (chestnut pumpkin) in Hebrew [3].

A technology of cleaning and disinfecting fresh-harvested produce in hot water (45–62 °C) over brushes for a very short time (15–25 s) is used commercially in Israel [5]. This technology is applied to several fruits and vegetables to reduce decay development and maintain fresh produce quality through prolonged storage and shelf life.

Very little information is available about the storage life of sweet acorn squash. Therefore, the goal of this research was to determine the best storage temperature for acorn squash fruits and to evaluate the use of hot water rinsing and brushing technology to clean and disinfect the fruits before prolonged storage for 3.5 months.

2. Materials and Methods

2.1 Plant Materials

Acorn squash fruits (*Cucurbita pepo* L. cv. Or) (Origene Seeds, Ltd., Giv'aat Brener, Israel), uniform in color and size with no defects, were harvested with a clipper from a commercial field in the central region of Israel. They were selected according to commercial maturity indexes (initial color ~125 Hue°; unit weight about 600–700 g).

2.2 Experiments and Treatments

In a marketing simulation in 2016 dry-brushed fruits were kept at 10, 15 or 20 °C, and 95, 95 or 70% RH, respectively for 2.5 month, followed by 3 days at 20 °C. In 2017, in light of preliminary results, the fruits were subjected to hot water rinsing and brushing (HWRB) treatment at 54 ± 1 °C for 15 s as described by Fallik [6] and kept at 15 or 20 °C for 3.5 months, and then for 3 days at 20 °C. Untreated fruits and those treated by tap water rinsing and brushing (TWRB) served as controls.

2.3 Quality Evaluation

Fruit quality was evaluated after 3.5 months at 15 or 20 °C followed by 3 days at 20 °C as follows: Weight loss percentage was calculated from the weights of 10 fruits before and after storage. Total soluble solids (TSS) contents were measured in 10 fruits by removing and squeezing a segment of flesh (from the peel to the seedbed) onto a digital refractometer (Atago, Tokyo, Japan); the same 10 fruits were tested for color, firmness and weight loss, Fruit epidermal color was evaluated with a Minolta Chroma Meter (Minolta, Ramsey, NJ, USA) that was calibrated against a white standard tile; two sides of each of 10 fruits were measured near the equator, and the results expressed as Hue angle (h°). Firmness was measured in newtons (N) (C-peak mode) with a motorized Chatillon penetrometer equipped with a 6-mm conic plunger (John Chatillon & Sons, New Gardens, NY, USA); each measurement was applied on opposite sides of each of 10 fruits, near the equator. Decay was expressed as the percentage of fruits with visible fungal mycelia on the peel.

2.4 Statistical Analysis

Three experiments/harvests were conducted each year. Each treatment consisted of 2 cartons, each consisted 14–15 fruits per carton (28–30 fruits per treatment), two repetitions per treatment. All data were subjected to one- or two-way statistical analysis at P = 0.05 with the JMP-11 Statistical Analysis Software Program (SAS Institute, Cary, NC, USA).

3. Results

3.1 Experiments in 2016

The best storage temperature, based on weight loss, decay incidence and color quality parameters after 2.5 months storage, was found to be 15 °C (Table 1). At 10 °C, although fruits lost less weight and were much greener, decay incidence was significantly higher than in the other treatments. At 20 °C, fruits lost significantly more weight and their color was greenish/orange, but decay incidence was similar to that at 15 °C (Table 1).

Table 1. The influence of storage temperature on Acorn squash quality parameters after 2.5 months in cold storage plus 3 days at 20°C.

Storage Temperature	Weight Loss (%)	Decay (%)	Color (Hue°)
10 °C	6.4 b [z]	55.3 a	103 a

15 °C	6.9 b	6.3 b	90 b
20 °C	13.4 a	8.3 b	78 c

z Values in the same column followed by the same letter are not significantly different at P = 0.05 according to Fisher's least significant difference test.

3.2 Experiments in 2017

Storing the fruits at 15 °C for 3.5 months significantly maintained fruit quality, based on firmness, decay development and color (Table 2, Figure 1). At this storage temperature, fruits were significantly firmer, showed less decay incidence, and were significantly greener than those in other treatments. TSS was not affected by storage temperature (Table 2; mean value at each temperature). Use of HWRB maintained significantly better fruit quality than that of untreated controls and TWRB-treated fruits (Table 2; mean value in each treatment; Figure 1). HWRB-treated fruits were very firm, had significantly less decay, and were significantly greener than untreated controls and TWRB-treated fruits. No significant differences were observed in TSS content although it was higher in HWRB-treated fruits (Table 2). No interaction between storage temperature and wash treatment was observed (Table 2).

Table 2. The influence of tap water rinsing and brushing (TWRB) and hot water rinsing and brushing (HWRB = 54 ± 1°C for 15 s) on Acorn squash quality parameters after 3.5 months at 15 or 20°C plus 3 days at 20°C.

Treatment-Wash	Temperature	Firmness (N)	Decay (%)	TSS (%)	Color (Hue°)
Control [z]	15 °C	83 b *	7.0 b	9.2 a	77.0 bc
TWRB [y]	15 °C	86 ab	3.7 cd	9.5 a	79.0 bc
HWRB [x]	15 °C	102 a	0.7 e	9.9 a	91.3 a
Control	20 °C	66 c	10.0 a	9.1 a	73.0 c
TWRB	20 °C	78 bc	4.7 bc	9.0 a	73.0 c
HWRB	20 °C	89 ab	1.7 de	9.8 a	80.7 b
LSD [w]	**7.31**	**1.21**	**1.96**	**3.34**	
		Mean values at each temperature			
15 °C		90.6 a	3.8 b	9.5 a	82.4 a
20 °C		77.8 b	5.4 a	9.3 a	75.6 b
LSD		**4.21**	**0.70**	**1.13**	**1.93**
		Mean values in each treatment			
Control		74.8 b	8.5 c	9.1 a	75.0 b
TWRB		82.3 ab	4.2 b	9.3 a	76.0 b
HWRB		95.3 a	1.2 c	9.9 a	86.0 a
LSD		5.16	0.86	1.39	2.36
		Table of Variance (F-value)			
Treatment (Tr)		*	*	NS	***

Temperature (Te)		**	***	NS	**
Tr × Te		NS	NS	NS	NS

*z Untreated control; y Tap water rinsing and brushing for 15 s; x Hot water rinsing and brushing for 15 s; w LSD, Least significance difference at $\alpha = 0.05$; v Means within columns followed by the same letter are not significantly different at $P \leq 0.05$, based on the least significant difference test; ***, **, *, NS indicate statistical significance at $P \leq 0.001$, 0.01, and 0.05, and not significant, respectively.*

Figure 1. The influence of storage temperature and water treatment on Acorn squash quality after 3.5 month at 15 or 20°C plus 3 days at 20°C (TWRB = tap water rinsing and brushing; HWRB = hot water rinsing and brushing at 54 ± 1°C for 15 s).

4. Discussion

Acorn squash has characteristic inedible, hard, thin skin, and firm flesh. The flesh is very sweet with a nut-like flavor after baking, microwaving, sautéing, or steaming. This squash is an excellent source of nutrients, including carotenoids, ascorbic acid, and vitamin C [7]. It is best known as a source of carotenoids—primarily b-carotene—and lutein, which are beneficial, respectively, as a provitamin A compound and for general health [8]. Market opportunities for growers have extended into the fall/

winter season and Acorn squash occupies an important late-season niche in Israel, therefore it is anticipated that this market has scope for increased growth in Israel and elsewhere [9].

To the best of our knowledge, there is very little published information about the optimal temperature for prolonged storage of Acorn squash. Acorn squash is considered a winter squash that can be stored at 10–12 °C for 2–3 months, while lower temperatures cause chilling injuries, and those above 20 °C increase weight loss and decay incidence [10]. In the present study 15 °C was found the best temperature to keep the fruit for 3.5 months. However, to keep the fruit at 15 °C for several months without affecting its quality, it is necessary to use HWRB treatment, whose beneficial effects on fresh-harvested produce were reported by Fallik [5,6]. HWRB treatments can remove fungal pathogens from the fruit surface through the brushing effect, and the natural wax platelets could be melted and smoothed to seal stomatal openings or invisible surface cracks, thereby reducing decay development and water loss and, in turn, increase fruit firmness (Tables 1 and 2) [5]. The lower decay incidence in HWRB-treated Acorn squash could also be attributed to the induction of pathogenesis-related proteins and the accumulation of enzymes such as chitinase and β-1,3-glucanase, which hydrolyze the fungal cell walls and inactivate the pathogens [5]. This treatment also was reported to delay fruit ripening, which may partially account for the delayed color development of the HWRB-treated fruits (Tables 1 and 2; Figure 1).

In conclusion, a prestorage HWRB treatment at 52°C for about 15 s, followed by storage at 15°C can maintain Acorn squash quality and marketability for several months. However, more research is needed to extend Acorn squash storability beyond 3.5 months, by evaluating new varieties and other prestorage treatments such as plant-growth regulators that delay chlorophyll degradation during prolonged storage, or by using edible coating materials.

References

1. Paris, H.S. A proposed subspecific classification for Cucurbita pepo. *Phytologia* **1986**, *61*, 133–138.
2. Paris, H. History of the cultivar-groups of Cucurbita pepo. *Hortic. Rev.* **2001**, *25*, 71–170.
3. Paris, H.; Godinger, D. Sweet acorn squash, a new vegetable on the Israeli market. *Acta Hortic.* **2016**, *1127*, 541–546. [CrossRef]
4. Paris, H.S.; Omer, S. An array of small pumpkins and squash: Now on the Israeli market, too. *Sade Ve-Yaraq* **2008**, *4*, 69–70. (In Hebrew)
5. Fallik, E. Prestorage hot water treatments (immersion, rinsing and brushing). *Postharvest Biol. Technol.* **2004**, *32*, 125–134. [CrossRef]
6. Fallik, E. Hot water treatments of fruits and vegetables for postharvest storage. *Hortic. Rev.* **2010**, *38*, 191–212.
7. Azevedo-Meleiro, C.H. Rodriguez-Amaya, D.B. Qualitative and quantitative differences in carotenoid composition among Cucurbita moschata, Cucurbita

maxima, and Cucurbita pepo. *J. Agric. Food Chem.* **2007**, *55*, 4027–4033. [CrossRef] [PubMed]

8. Wyatt, L.E.; Strickler, S.R.; Mueller, L.A.; Mazourek, M. An acorn squash (*Cucurbita pepo* ssp. ovifera) fruit and seed transcriptome as a resource for the study of fruit traits in *Cucurbita. Hortic Res.* **2015**, *2*, 14070. [CrossRef] [PubMed]

9. Slosar, M.; Mezeyova, I.; Hegedusova, A.; Hegedus, O. Quantitative and qualitative parameters in Acorn squash cultivar in the conditions of the Slovak Republic. *Potravin. Slovak J. Food Sci.* **2018**, *12*, 91–98.

10. Gibe, A.J.G.; Yoon, K.S.; Park, K.S.; Lee, J.W. Influence of curing on the quality of winter squash (*Cucurbita maxima* Duch. *Var. ebis Takii*) stored at different temperatures. *Asia Life Sci.* **2008**, *17*, 325–336.

Chapter-7

Postharvest Techniques to Prevent the Incidence of Botrytis Mold of 'BRS Vitoria' Seedless Grape under Cold Storage

1. Introduction

Released in 2012, the new hybrid seedless table grape 'BRS Vitoria' is recommended for cultivation in tropical and subtropical areas. This cultivar has good development and production, and is tolerant to downy mildew (*Plasmopara viticola*), one of the major grape diseases in humid areas. Due to its excellent flavor and firmness, this cultivar is a good option not only for the internal market, but also for export [1]. Therefore, it is necessary to develop techniques that allow postharvest conservation of this grape in cold storage for long periods.

The fungus *Botrytis cinerea*, which causes gray mold, is considered one of the most damaging postharvest pathogens to the quality of table grapes during storage and transport over long distances.

The control of this fungus is particularly important during refrigerated storage, as it also develops at low temperatures (−0.5 °C) and spreads rapidly through the berry clusters [2–4].

Sulfur dioxide (SO_2) is the main fungicide treatment used to retard the growth of this fungus in refrigerated chambers, and the purpose of its use is to inhibit fungus development and to allow the storage and transport of table grapes for long periods of time [5]. There are two main concepts for packaging grapes in prolonged storage: packing with SO_2 pads and fumigation of refrigeration chambers by repeated application of this gas [6,7]. Despite its effectiveness, there are restrictions for the fumigation of SO_2, as this may compromise fruit flavor, cause damage to the berries, and result in excessive sulfite residues [8].

Thus, the use of SO_2-generating pads was developed because they provide good and efficient control and lower risk than fumigation. Depending on the grape cultivar, different types of pads were developed, such as slow release and dual phase release, with different concentrations of SO_2 [9]. The SO_2-generating pads contain sodium metabisulfite, which when in contact with moisture inside the grape packaging, reacts by releasing SO_2 gas [10].

The choice of SO_2-generating pads should be judicious, in order to maintain the quality of the grape to its final destination, and to that end, the level of the active ingredient should be specific for each table grape cultivar. The main import markets for fresh grapes, such as the European Community and the United States, have established levels of tolerance to the use of SO_2 in postharvest management, aiming at greater protection of the consumer and also of the environment, since a high gas concentration and/or residue may be harmful to man and the environment [11].

Therefore, the evaluation of techniques to avoid grape losses during postharvest storage are needed to maintain quality and profit. In this context, the objective of this work was to evaluate different types of SO_2 generator pads that prevent the incidence of gray mold of 'BRS Vitoria' seedless grape, as well to avoid other grape injuries during cold storage.

2. Materials and Methods

Grape bunches were obtained from a commercial field of 'BRS Vitoria' seedless table grape grafted on 'IAC 766' rootstock, located at Marialva, state of Parana (PR) (South Brazil) (23°29′ S, 51°47′ W, elevation 570 m). Samples were collected from regular crops in 2018. According to the Köppen classification, the climate is type Cfa, i.e., subtropical with an average temperature in the coldest month below 18 °C, and average temperature in the warmest month above 22 °C. The maximum temperature is 31 °C, and the average annual rainfall is 1596 mm, with a tendency for concentrated rainfall in summer. The field was selected because of its history of gray mold incidence.

Botrytis cinerea, used in this study, was isolated from infected grapes showing typical gray mold symptoms, purified and identified morphologically and molecularly [12]. The isolates were maintained on potato dextrose agar (PDA) slants and stored at 4 °C for further use. Fungal conidia were harvested from 2-week-old PDA cultures of Botrytis cinerea grown at 23 ± 1°C. A volume of 5 mL of distilled water, containing 0.05% (*v/v*) Tween 80, was added to a Petri plate culture. The conidia were gently dislodged from the surface with a distilled glass rod, and suspensions were filtered through three layers of cheesecloth to remove any adhering mycelia. The suspensions were diluted with sterile water and the concentration was determined with a hemacytometer. Further dilutions with sterile water were made to obtain the desired conidial concentrations. *Botrytis cinerea* suspension (10^6 conidia mL^{-1}) was used for grape inoculation.

Grapes were harvested at full maturity when soluble solids content reached around 16°Brix, and then, bunches were subjected to the following treatments: (a) control; (b) SO_2 slow release pad; (c) SO_2 dual release pad; (d) SO_2 dual release–fast reduced pad; (e) SO_2 slow release pad with grapes inoculated with *Botrytis cinerea*; (f) SO_2 dual release pad with grapes inoculated with *Botrytis cinerea*; (g) SO_2 dual release–fast reduced pad with grapes inoculated with *Botrytis cinerea*. The inoculation was carried out by spraying a conidial suspension (10^6 conidia mL^{-1}). A completely randomized experimental design was used with seven treatments and four replicates, with five bunches per plot. Before packing, bunches were subjected to forced air precooling, cleaned, and the damaged berries were removed. Then, they were standardized according to their appearance and mass (~0.5 kg), and arranged individually in a plastic clamshell of 0.5 kg capacity, measuring 20 × 10 cm. The process of packaging the grapes followed several steps: arrangement of plastic micro perforated liner films in 4.5 kg-capacity carton boxes measuring 50 × 30 × 10 cm; deposition of a sheet of moisture-absorbing paper, measuring 37 × 28 cm on the bottom of the liner; placement of plastic clamshells with grapes; arrangement of the SO_2 pad on top; and sealing of liner. The cartons boxes were then placed in a cold chamber at 2°C with high relative humidity for 50 days followed by one week of room temperature at 22 ± 2°C. For treatments with SO_2, one generator pad measuring 37 × 28 cm per box provided with slow release, dual release, or dual release–fast reduced phases of sodium metabisulfite ($Na_2S_2O_5$) (Uvasys®, Cape Town, South Africa) was used, containing 3.85 g; 4.50 g, and 4.25 g of active ingredient, respectively. The dual release–fast reduced pad was designed for SO_2-sensitive cultivars, since it releases 60% of the active ingredient during the first phase, and 40% during the subsequent fast reduced phase.

The incidence of gray mold on grapes was evaluated at 50 days after the beginning of cold storage and at 7 days at room temperature after the end of cold storage. The disease incidence was then obtained by the formula: disease incidence (%) = (number of infected bunches/total number of bunches) × 100 [12].

Stem browning development was measured by using the following scoring system: (1) fresh and green; (2) some light browning; (3) significant browning; and (4) severe browning, calculated by the weighted average of the scale value and number of bunches at each level, ranging between 1 and 4 [13] (Figure 1). The shattered berries incidence of was determined by counting the separated berries from the bunch stem inside the clamshell, and were expressed as a percentage of the total number of berries.

Figure 1. Stem browning scores of the 'BRS Vitoria' table grapes. 1: fresh and green; 2: light browning; 3: moderate browning; and 4: severe browning.

The bunch mass loss (%) during postharvest storage was determined by periodic weighing, and calculated by dividing the mass change during storage by the original mass: mass loss (%) = [(mi − ms)/mi] × 100, where mi = initial mass and ms = mass at examining time [14]. The berry firmness or maximum compression force was performed with a texture analyzer TA.XT*Plus* (Stable Micro Systems, Surrey, UK), analyzing the equatorial position of 10 berries with pedicels per plot. Each berry was placed on the base of the equipment and compressed using a cylindrical probe with a diameter of 35 mm parallel to the base. A constant force of 0.05 N at a speed of 1.0 mm s^{-1}

was applied to deform the berry to 20% of its equatorial diameter. The berry firmness (N) was then determined [15].

Berry color was analyzed using a colorimeter Minolta® CR-10 to obtain the following variables from the equatorial portion of berries (n = 2 measurements per berry): L^* (lightness), C^* (chroma), and h° (hue angle) [16]. Lightness values may range from 0 (black) to 100 (white). Chroma indicates the purity or intensity of color, the distance from gray (achromatic) toward a pure chromatic color and is calculated from the a^* and b^* values of the CIELab scale system (International Comission on Illumination, Vienna, Austria), starting from zero for a completely neutral color, and does not have an arbitrary end, but intensity increases with magnitude. Hue refers to the color wheel and is measured in angles; green, yellow, and red correspond to 180°, 90°, and 0°, respectively [17–19].

For chemical analysis, 15 berries were collected from each plot. The juice was used to determine soluble solids (SS) and titratable acidity (TA). SS was determined with a digital refractometer (Krüss, Hamburg, Germany) at 20 °C, and the results were expressed in °Brix. The pH of the juice was recorded using a Jenway 3510 bench pH meter (Cole-Parmer, Staffordshire, UK) and then TA was determined by potentiometric titration with NaOH 0.1 N up to pH 8.2, using 10 mL of diluted juice in 40 mL distilled H_2O, and the results were expressed in % of tartaric acid [20]. The grape physicochemical analysis was performed at 50 days of cold storage and 7 days of room temperature. All data were subjected to analysis of variance using Sisvar software (UFLA, Lavras, Brazil).

Mean values of treatments were compared by using Scott Knott's test and judged at $p \leq 0.05$ levels.

3. Results

3.1 Gray Mold Incidence (%)

After 50 days of cold storage, 'BRS Vitoria' table grapes submitted to treatments with SO_2-generating pads presented a lower incidence of gray mold in relation to the control, and in the treatments with the dual release and fast-reduced release pads, no symptoms of the disease were observed, even when the fruit were inoculated with pathogen (Table 1).

Table 1. Gray Mold Incidence (%) of 'BRS Vitoria' seedless table grapes at 50 days of cold storage and 7 days of room temperature under different SO2 pad treatments.

	Gray Mold Incidence (%)	
Treatments	**At 50 Days in Cold Storage**	**At 7 Days in Room Temperature**
Control	7.9 ± 1.4 a [z]	13.0 ± 3.0 a
SO_2 slow release pad	1.3 ± 1.3 b	13.4 ± 8.4 a
SO_2 dual release pad	0.1 ± 0.2 c	6.6 ± 2.3 b
SO_2 dual release-fast reduced pad	0.0 ± 0.0 c	3.3 ± 2.0 b
SO_2 slow release pad + Bo [y]	1.9 ± 0.9 b	17.2 ± 3.8 a
SO_2 dual release pad + Bo	0.3 ± 0.5 c	4.7 ± 3.2 b

SO_2 dual release–fast pad reduced + Bo	0.0 ± 0.0 c	2.5 ± 3.3 b
F value	34.5 ** X	5.6 **

z Means within columns followed by the same letters are not significantly different by Scott Knott's test (p 0.05).

*y Bo: Botrytis inoculation. x **: F value significant at 1%.*

In the same way, these treatments presented better performance in fungus control after 7 days at room temperature, where the highest incidence of gray mold was observed in treatments with slow release pads, with or without inoculation (13.4% and 17.2%, respectively), and in the control (13%) Table 1). Thus, it was verified that the slow release pad was less efficient to avoid the occurrence of gray mold in the 'BRS Vitoria' grape stored in cold chamber.

3.2 Shattered Berries and Stem Browning

It was observed that all treatments with SO_2 generator pads showed a lower percentage of shattered berries (1.5% to 3.5%) at 50 days after refrigerated storage when compared with control (6.1%). At 7 days after room temperature there was no difference among the treatments, and the percentage of shattered berries varied from 3.1% to 7.2% (Table 2).

Table 2. Shattered berries (%) and stem browning scores of 'BRS Vitoria' seedless table grapes at 50 days in cold storage and 7 days of room temperature under different SO_2 pad treatments.

Treatments	Shattered Berries (%)	Stem Browning Scores z	Shattered Berries (%)	Stem Browning Scores z
	At 50 Days in Cold Storage		At 7 Days in Room Temperature	
Control	6.1 2.8 a y	1.8 0.1	4.9 2.1	2.8 0.5
SO_2 slow release pad	2.9 1.8 b	1.3 0.3	4.8 2.2	2.5 0.4
SO_2 dual release pad	1.6 1.3 b	1.3 0.3	3.7 3.5	2.5 0.2
SO_2 dual release–fast reduced pad	1.6 0.7 b	1.5 0.2	3.1 0.6	2.3 0.3
SO_2 slow release pad + Bo x	3.3 1.5 b	1.6 0.2	7.2 1.3	2.8 0.2
SO_2 dual release pad + Bo	3.5 0.9 b	1.3 0.2	5.0 1.1	2.2 0.2
SO_2 dual release–fast pad reduced + Bo	1.5 0.3 b	1.4 0.1	3.4 0.6	2.7 0.4
F value	3.4 * w	2.14 NS	1.6 NS	1.40 NS

*z Stem Browning Scoring System: 1 = fresh and green; 2 = light browning; 3 = moderate browning; and 4 = severe browning. y Means within columns followed by the same letters are not significantly different by Scott Knott's test ($p \leq 0.05$). x Bo: Botrytis inoculation. w *: F value significant at 5%. NS: not significant.*

The browning was not affected by the treatments in the evaluated periods. However, at 50 days after refrigerated storage, the scores were closer to 1 and 2, indicating fresh and green and browning, and after 7 days at room temperature, the scores were close to 3, indicating moderate browning (Table 2).

3.3 Physical Properties for Quality Measurements

In the treatments with SO_2 slow release pads, the mass loss was lower (1.0%) than the other treatments (1.8% to 2.4%) at 50 days after refrigerated storage. On the other hand, at 7 days at room temperature, it was observed that the treatments with SO_2 dual release and dual release–fast reduced pads had lower mass loss, even when inoculated with *Botrytis* (0.3% and 0.6%) (Table 3).

Table 3. Mass loss and firmness of 'BRS Vitoria' seedless table grapes at 50 days of cold storage and 7 days of room temperature under different SO_2 pad treatments.

	Mass Loss (%)	Firmness (N)	Mass Loss (%)	Firmness (N)
Treatments	**At 50 Days in Cold Storage**		**At 7 Days in Room Temperature**	
Control	1.9 ± 0.2 a z	11.9 ± 1.2	1.7 ± 0.2 a	12.0 ± 1.1
SO_2 slow release pad	1.0 ± 0.2 b	12.7 ± 0.4	1.9 ± 0.7 a	12.5 ± 1.5
SO_2 dual release pad	2.1 ± 0.3 a	10.1 ± 0.6	1.1 ± 0.1 b	11.3 ± 1.2
SO_2 dual release–fast reduced pad	1.8 ± 0.3 a	11.4 ± 1.2	0.9 ± 0.2 b	11.5 ± 1.1
SO_2 slow release pad + Bo y	2.2 ± 1.2 a	12.5 ± 2.4	1.6 ± 0.2 a	14.9 ± 1.7
SO_2 dual release pad + Bo	2.1 ± 0.3 a	13.4 ± 1.1	0.6 ± 0.3 c	13.3 ± 1.1
SO_2 dual release–fast pad reduced + Bo	2.4 ± 0.1 a	11.1 ± 0.4	0.3 ± 0.1 c	12.5 ± 1.6
F value	2.7 * x	2.3 NS	12.4 **	2.5 NS

[z]*Means within columns followed by the same letters are not significantly different by Scott Knott's test ($p \leq 0.05$).*

[y]*Bo: Botrytis inoculation. x **, *: significant at 1% and 5%, respectively. NS: not significant.*

There was no difference between the treatments in relation to firmness of the berries 50 days after refrigerated storage and 7 days after ambient temperature; the averages varied from 10.1 N to 12.7 N, and 11.3 to 14.9 N, respectively (Table 3).

3.4 Chemical Properties for Quality Measurements

Regarding the color attributes of grape berries, no differences were observed among treatments at 50 days after refrigerated storage. After 7 days at room temperature, grapes treated with SO_2 dual release–fast reduced pads presented higher means of L^*, indicating a darker coloration, and when inoculated with *Botrytis*, lower means of C^*. In the same period, a lower h° (57.3) was observed, with saturation between orange and yellow for treatment with SO_2 dual release pads with inoculation (Table 4).

There was no difference between treatments regarding the pH of the grapes in the evaluated periods (Table 5). The grapes treated with dual release–fast reduced and slow release pads with inoculation of *Botrytis* showed higher TA at 50 days after refrigerated storage. In the same period, the highest SS means were observed in treatments with slow release, dual release, and dual release–fast reduced pads with inoculation. After 7 days at room temperature, all treatments with SO_2 pads inoculated with *Botrytis* presented higher levels of SS. The SS/AT ratio was lower in treatments with slow release pads with *Botrytis* inoculation, dual release, and dual release–fast reduced pads 50 days after refrigerated storage, but at 7 days at room temperature, slow release pads and all treatments inoculated with *Botrytis* showed higher means for this characteristic (Table 5).

Table 4. Luminosity (L*), chroma (C*), and hue angle (h*) of 'BRS Vitoria' seedless table grape at 50 days of cold storage and 7 days of room temperature under different SO2 pad treatments.

Treatments	L*	C*	h°	L*	C*	h°
	At 50 Days of Cold Storage			At 7 Days of Room Temperature		
Control	20.0 ± 0.1	4.1 ± 0.8	117.6 ± 19.3	19.9 ± 0.5 b [z]	2.8 ± 0.1 a	101.1 ± 9.5 a
SO_2 slow release pad	20.0 ± 0.2	3.4 ± 0.4	124.8 ± 7.8	20.0 ± 0.6 b	2.6 ± 0.1 a	92.3 ± 4.2 a
SO_2 dual release pad	19.5 ± 0.3	3.2 ± 0.2	118.5 ± 7.6	20.0 ± 0.5 b	2.7 ± 0.2 a	92.4 ± 11.0 a
SO_2 dual release-fast reduced pad	20.2 ± 0.3	3.2 ± 0.1	110.0 ± 14.5	20.1 ± 0.3 b	2.2 ± 0.2 b	95.4 ± 15.5 a
SO_2 slow release pad + Bo [y]	19.9 ± 0.7	3.2 ± 0.3	111.3 ± 4.5	19.8 ± 0.4 b	2.6 ± 0.4 a	97.0 ± 16.1 a
SO_2 dual release pad + Bo	19.9 ± 0.5	3.2 ± 0.4	116.0 ± 7.4	20.2 ± 0.6 b	3.0 ± 0.1 a	57.3 ± 8.2 b
SO_2 dual release-fast pad reduced + Bo	19.9 ± 0.2	3.3 ± 0.4	124.7 ± 10.5	22.2 ± 0.4 a	2.1 ± 0.4 b	82.3 ± 1.4 a
F value	0.9 [NS x]	1.9 [NS]	0.8 [NS]	8.2 **	5.6 **	5.8 **

*[z]Means within columns followed by the same letters are not significantly different by Scott Knott's test ($p \leq 0.05$). y Bo: Botrytis inoculation. x **: significant at 5%. [NS]: not significant. Table 5. pH, soluble solids*

(SS), titratable acidity (TA), SS/AT of 'BRS Vitoria' seedless table grape at 50 days of cold storage, and 7 days of room temperature under different SO2 pad treatments.

Table 5. pH, soluble solids (SS), titratable acidity (TA), SS/AT of 'BRS Vitoria' seedless table grape at 50 days of cold storage, and 7 days of room temperature under different SO_2 pad treatments.

Treatments	pH	Soluble Solids—SS (°Brix)	Titratable Acidity—TA (%)	SS/TA	pH	Soluble Solids—SS (°Brix)	Titratable Acidity—TA (%)	SS/TA
	At 50 Days of Cold Storage				At 7 Days of Room Temperature			
Control	4.0 ±0.1	17.2 ±0.8 b [z]	0.7 ±0.0 b	24.1 ±1.7 a	4.0 ±0.1	16.8 ±0.6 b	0.7 ±0.0	25.6 ±1.5 b
SO_2 slow release pad	4.1 ±0.2	17.6 ±0.5 a	0.7 ±0.0 b	24.7 ±1.4 a	4.0 ±0.1	16.8 ±0.8 b	0.6 ±0.0	28.9 ±3.5 a
SO_2 dual release pad	3.9 ±0.1	17.0 ±0.4 b	0.8 ±0.0 b	22.9 ±1.5 b	4.1 ±0.2	16.1 ±0.6 b	0.6 ±0.0	25.2 ±0.6 b
SO_2 dual release-fast reduced pad	4.1 ±0.1	17.4 ±0.3 b	0.8 ±0.0 a	22.1 ±1.0 b	4.1 ±0.2	16.3 ±0.9 b	0.6 ±0.0	25.8 ±3.3 b
SO_2 slow release pad + Bo [y]	4.0 ±0.1	16.8 ±0.3 b	0.8 ±0 a	20.7 ±1.1 b	4.1 ±0.1	17.7 ±0.3 a	0.6 ±0.0	28.8 ±0.3 a
SO_2 dual release pad + Bo	4.1 ±0.2	18.0 ±0.4 a	0.7 ±0.0 b	24.5 ±1.4 a	4.0 ±0.1	17.5 ±0.3 a	0.6 ±0.0	29.0 ±1.9 a
SO_2 dual release-fast pad reduced + Bo	4.1 ±0.2	17.9 ±0.5 a	0.8 ±0.0 b	23.5 ±0.4 a	4.2 ±0.1	18.5 ±1.2 a	0.6 ±0.0	29.9 ±2.0 a
F value	0.8 [NS x]	2.7 *	4.7 **	4.2 **	0.8 [NS]	3.4 *	1.1 [NS]	2.4 *

*z Means within columns followed by the same letters are not significantly different by Scott Knott's test ($p \leq 0.05$). y Bo: Botrytis inoculation. x **, *: significant at 1% and 5%, respectively. NS: not significant.*

4. Discussion

The fungus *Botrytis cinerea*, which causes gray mold, is considered one of the most harmful postharvest pathogens to table grapes, especially during transport and storage. The use of SO_2-generating pads is one of the main fungus control methods, especially when grapes are stored in a cold room for prolonged periods [10], which is necessary if the objective is to export 'BRS Vitoria' seedless grape, or even to store it for the domestic market.

It was verified that the SO_2 generator pads reduced the incidence of gray mold at 50 days after refrigerated storage, with emphasis on the dual release–fast reduced pads, with no symptoms of the disease, even when the pathogen was inoculated. The grapes treated with dual release pads also presented better results in the presence of the fungus at 7 days at room temperature, and were more promising than the SO_2 slow release pad for 'BRS Vitoria' table grapes, which were not efficient in control of the disease. In addition, 'BRS Vitoria' seedless grape was not sensitive to SO_2 emitted in the fast phase of the dual release pads tested, nor to other disorders, such as hairline, bleaching, and cracking. The use of clamshells for the refrigerated storage of this grape cultivar has been shown to be suitable for use in combination with the SO_2 pads, because it allows a sufficient ventilation area for good circulation of the active ingredient in the package.

In another study, SO_2 generator pads contributed to the reduction of gray mold in 'BRS Vitoria' seedless grapes after 30 days of refrigerated storage, with a minimum reduction of 80% [21], but lower than observed in this trial, since the amount of active ingredient was smaller. In addition, the use of SO_2 pads in combination with packaging methods using macroperforated liners had high disease control of 'Red Globe' table grapes stored for four months [22]. The use of suitable packaging combined with the conscious and efficient use of SO_2 pads and storage at low temperatures (0–2 °C, 90–95% RH) may contribute to the maintenance of the postharvest quality of table grapes up to 50 days [23].

The choice of SO_2 pads should be judicious to maintain the quality of each table grape cultivar to its final destination, so the level of the active ingredient should be adequate, in order to not injure the fruits or to damage their flavor. The main import markets for fresh grapes, such as the European Community and the United States, have established levels of tolerance to the use of SO_2 in postharvest management, aiming at greater protection of the consumers and of the environment [11].

The mechanism used by SO_2 pads occurs due to the presence of moisture in the packaging of the grapes, which is absorbed by the sheets and that reacts with the active ingredient, sodium metabisulfite ($Na_2S_2O_5$), resulting in the release of the gas to the external environment. Slow release SO_2 pads have lower permeability and emit low SO_2 concentrations over a long period of up to 60 days. The dual release pads contain the $Na_2S_2O_5$ enclosed between two plastic sheets, a slow release part and a fast release part, which provides different permeabilities, so the gas release takes place in a way specific to the part. The fast part releases SO_2, which peaks after 24 h, and then decreases by one week. The slow release of the dual release pads emits SO_2 continuously, ensuring the presence of the gas for extended periods [9,10]. The use of the dual release pad in this work proved to be more efficient than the slow release pad, which leads to the belief that the SO_2 fast release phase in the first 48 h of the dual release pads is necessary for the conservation of healthy 'BRS Vitoria' grapes, even though the pathogen had been inoculated artificially.

Bleaching and cracking on the surface of berries, and in some cases, stem browning, are unfavorable characteristics that impair the quality of the table grapes preserved in the cold, and can develop through the indiscriminate use of SO_2 pads due to the release of excessive levels of the active ingredient [22,24]. In addition to the browning and dehydration of the rachis, the loss of water and the degradation of the berries are factors that also decrease the postharvest life of table grapes during storage and transportation [25,26], and should, therefore, be considered when new techniques or materials for packaging are assessed. In this study, the SO_2 pads did not affect the stem browning in the evaluated periods. The stem browning scores indicated that the rachis color was between 'fresh and green' and 'light brown' after 50 days of refrigerated storage. Other authors have observed that SO_2 pads maintained a better fresh appearance by preventing browning in 'Thompson Seedless' [27] and 'Red Globe' table grapes [28]. The green and fresh color of the rachis is more attractive to the consumer than when darker [7]; therefore, with a view to commercialization, rachis color is an important parameter to be evaluated.

Shattered and softening berries, and mass loss, are some of the consequences caused by the incidence of *Botrytis cinerea* in table grapes [2,3]. In this trial, the grapes packed and treated with the SO_2 pads showed lower shattered berry incidence after the 50-day cold storage period, as well as a lower incidence of gray mold. In the same way, the effectiveness of gray mold control by the dual release and dual release–fast reduced pads, especially when berries were inoculated with *Botrytis*, minimized the mass loss of the grapes at 7 days after room temperature. Therefore, the use of SO_2 pads with high efficiency in the control of *Botrytis cinerea* can contribute to the maintenance of these characteristics and, consequently, maintain 'BRS Vitoria' table grapes in long periods of storage, especially when the objective is to preserve grapes over 30 days.

The analysis of the quality profile of the table grapes after the refrigerated storage period, especially in conjunction with techniques such as the use of SO_2 pads, is extremely important since any change in the taste or appearance of the fruit may interfere with the commercialization process. The results showed that the effects on the main quality parameters of 'BRS Vitoria' seedless table grapes when treated with SO_2 pads were not considerably detrimental. According to international marketing regulations, the minimum SS contents may vary according to the grape cultivar from 14.0 to 17.5°Brix and the SS/AT ratio must be equal to or higher than 20 [29]; values within what were observed in this work.

5. Conclusions

The dual release pads were more effective at preventing the incidence of gray mold and mass loss in 'BRS Vitoria' seedless grapes than slow release pads in both storage periods. Incidence of shattered berries were lower when any type of SO_2 pad was used during cold storage, and no effects were observed on stem browning, firmness, and berry skin color of 'BRS Vitoria' grapes. In this context, conscientious and effective use of SO_2 pads for packaging grapes, especially the dual release pad, enabled the long-term maintenance of 'BRS Vitoria' table grapes under refrigerated storage conditions.

References

1. Maia, J.D.G.; Ritschel, P.; Camargo, U.A.; Souza, R.T.; Fajardo, V.T.; Naves, R.L.; Girardi, C. 'BRS Vitoria'— A novel seedless table grape cultivar exhibiting special flavor and tolerance to downy mildew (*Plasmopara viticola*). *Crop Breed. Appl. Technol.* **2014**, *14*, 204–206. [CrossRef]
2. Bulit, J.; Dubos, B. Botrytis bunch rot and blight. In *Compendium of Grape Diseases*; APS Press: St. Paul, Rockville, MD, USA, 1990; pp. 13–15.
3. Celik, M.; Kalpulov, T.; Zutahy, Y.; Ish-Shalom, B.; Lurie, S.; Lichter, A. Quantitative and qualitative analysis of Botrytis inoculated on table grapes by *q*PCR and antibodies. *Postharvest Biol. Technol.* **2009**, *52*, 235–239. [CrossRef]
4. Oliveira, C.E.V.; Magnani, M.; de Sales, C.V.; Pontes, A.L.D.S.; Campos-Takaki, G.M.; Stamford, T.C.M. Effects of postharvest treatment using chitosan from *Mucor circinelloides* on fungal pathogenicity and quality of table grapes during storage. *Food Microbiol.* **2014**, *44*, 211–219. [CrossRef] [PubMed]
5. Crisosto, C.H.; Smilanick, J.L.; Dokoozlian, N.K.; Luvisi, D.A. Maintaining table grape postharvest quality for long distant markets. In Proceedings of the International Symposium on Table Grape Production, Anaheim, CA, USA, 28–29 June 1994; pp. 195–199.
6. Karabulut, O.A.; Mlikota Gabler, F.; Mansour, M.; Smilanick, J.L. Postharvest ethanol and hot water treatments of table grapes to control gray mold. *Postharvest Biol. Technol.* **2004**, *34*, 169–177. [CrossRef]
7. Lichter, A.; Gabler, F.M.; Smilanick, J.L. Control of spoilage in table grapes. *Stewart Postharvest Rev.* **2006**. [CrossRef]
8. Lichter, A.; Zutahy, Y.; Kaplunov, T.; Shacham, Z.; Aharoni, N.; Lurie, S. The benefits of modified atmosphere of ethanol treated grapes. *Acta Hortic.* **2005**, *682*, 1739–1744. [CrossRef]
9. Fernandez-trujillo, J.P.; Obando-Ulloa, J.M.; Baró, R.; Martinez, J.A. Quality of two table grape cultivars treated with single or dual-phase release SO_2 generators. *J. Appl. Bot. Food Qual.* **2008**, *82*, 1–8.
10. Lichter, A.; Zutahy, Y.; Kaplunov, T.; Lurie, S. Evaluation of table grapes storage in boxes with sulfur dioxide-releasing pads with either an internal plastic liner or external wrap. *HortTechnology* **2008**, *18*, 206–214.
11. Pires, J.C.M.; Sousa, S.I.V.; Pereira, A.M.C.; Alvim Ferraz, M.C.M.; Martins, F.G. Management of air quality monitoring using principal component and cluster analysis—Part I: SO_2 and PM10. *Atmos. Environ.* **2008**, *42*, 1249–1260. [CrossRef]
12. Youssef, K.; Roberto, S.R. Applications of salt solutions before and after harvest affect the quality and incidence of postharvest gray mold of 'Italia' table grapes. *Postharvest Biol. Technol.* **2014**, *87*, 95–102. [CrossRef]

13. Mduduzi, E.K.N.; Mulugeta, A.D.; Umezuruik, L.O.; Chris, J.M. Performance of multi-packaging for table grapes based on airflow, cooling rates and fruit quality. *J. Food Eng.* **2012**, *116*, 613–621. [CrossRef]

14. Mattiuz, B.; Miguel, A.C.A.; Galati, V.C.; Nachtigal, J.C. Efeito da temperatura no armazenamento de uvas apirênicas minimamente processadas. *Rev. Bras. Frutic.* **2009**, *31*, 44–52. [CrossRef]

15. Lijavetzky, D.; Carbonell-Bejerano, P.; Grimplet, J.; Bravo, G.; Flores, P.; Fenoll, J.; Hellin, P.; Oliveros, J.C.; Martinez-Zapater, J.M. Berry flesh and skin ripening features in *Vitis vinifera* as assessed by transcriptional profiling. *PLoS ONE* **2012**, *7*, e39547. [CrossRef]

16. Koyama, R.; Assis, A.M.; Yamamoto, L.Y.; Borges, W.S.; Prudencio, S.H.; Roberto, S.R. Exogenous abscisic acid increases the anthocyanin concentration of berry and juice from 'Isabel' grapes (*Vitis labrusca* L.). *HortScience* **2014**, *49*, 460–464.

17. McGuire, R.G. Reporting of objective color measurements. *HortScience* **1992**, *27*, 1254–1255.

18. Lancaster, J.E.; Lister, C.; Reay, P.F.; Triggs, C.M. Influence of pigment composition on skin color in a wide range of fruits and vegetables. *J. Am. Soc. Hortic. Sci.* **1997**, *122*, 594–598.

19. Peppi, M.C.; Fidelibus, M.W.; Dokoozlian, N. Abscisic acid application timing and concentration affect firmness, pigmentation and color of 'Flame Seedless' grapes. *HortScience* **2006**, *41*, 1440–1445.

20. Youssef, K.; Roberto, S.R. Salt strategies to control Botrytis mold of 'Benitaka' table grapes and to maintain fruit quality during storage. *Postharvest Biol. Technol.* **2014**, *95*, 95–102. [CrossRef]

21. Youssef, K.; Roberto, S.; Chiarotti, F.; Koyama, R.; Hussain, I.; Souza, R. Control of Botrytis mold of the new seedless grape 'BRS Vitoria' during cold storage. *Sci. Hortic.* **2015**, *193*, 316–321. [CrossRef]

22. Zutahy, Y.; Lichter, A.; Kaplunov, T.; Lurie, S. Extended storage of 'Red Globe' grapes in modified SO_2 generating pads. *Postharvest Biol. Technol.* **2008**, 12–17. [CrossRef]

23. Zoffoli, J.P.; Latorre, B.; Naranjo, P. Hairline, a postharvest cracking disorder in table grapes induced by sulfur dioxide. *Postharvest Biol. Technol.* **2008**, *47*, 90–97. [CrossRef]

24. Lurie, S.; Pesis, E.; Gadiyeva, O.; Feygenberg, O.; Ben-Arie, R.; Kaplunov, T.; Zutahy, Y.; Lichter, A. Modified ethanol atmosphere to control decay of table grapes during storage. *Postharvest Biol. Pathol.* **2006**, *42*, 222–227. [CrossRef]

25. Zoffoli, J.P.; Latorre, B.A.; Naranjo, P. Preharvest applications of growth regulators and their effect on postharvest quality of table grapes during cold storage. *Postharvest Biol. Technol.* **2009**, *51*, 183–192. [CrossRef]

26. Meng, X.; Tian, S. Effects of preharvest application of antagonistic yeast combined with chitosan on decay and quality of harvested table grape fruit. *J. Sci. Food Agric.* **2009**, *89*, 1838–1842. [CrossRef]

27. Sölylemezoglu, G.; Agaoglu, Y.S. Research on the effect of grape guard during the cold storage of Thompson seedless cv. *Acta Hortic.* **1994**, *368*, 817–824. [CrossRef]
28. Ozkaya, O.; Dundar, O.; Ozdemir, A.E.; Dilbaz, R. Effects of different postharvest applications on Red Globe grapes storage. *Alatarım* **2005**, *4*, 44–50. (In Turkish with English Abstract)
29. Bleinroth, E.W. Determinação do ponto de colheita. In *Uva Para Exportação: Procedimentos De Colheita E* Pós-Colheita; EMBRAPA—SPI/FRUPEX: Brasília, Brazil, 1993; pp. 20–21.

Chapter-8

Fruit Stem-End Rot

1. Characterization of Stem-End Rot Causing Pathogens and Their Lifestyle

In recent years, there has been a rising demand for ripe and ready-to-eat fruit. However, as the fruit ripens, it becomes susceptible to various postharvest diseases [1]. Among them is the emergence of stem-end rot (SER) disease. SER occurs in various fruit, and particularly in tropical and subtropical fruit, including mango, avocado, citrus, mangosteen, carambola, and others. In mangoes, for example, SER is considered to be the second most severe disease worldwide, after anthracnose, caused by *Colletotrichum gloeosporioides* [2], while in dry areas, SER is the major postharvest pathogen. For

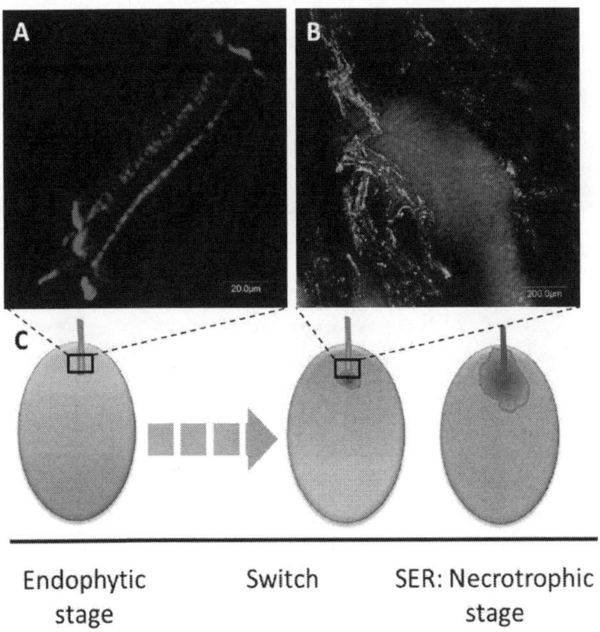

Figure 1. Stem-end rot development during the fruit ripening process. Confocal images of mango stem-end stained with aniline blue show (A) endophytic colonization of the phloem, and (B) necrotrophic colonization of ripe fruit. (C) Illustration of the development of stem-end rot during fruit ripening (adapted from Diskin et al. [6] with permission).

example, in Israel, SER caused 30–40% loss of harvested mango fruit during 2014 (Diskin et al., in press).

SER-causing pathogens penetrate to the stem through natural openings and wounds, mainly during inflorescence and flowering stages [3–5]. Those fungal pathogens live endophytically, mainly in the phloem but also in the xylem, and exist asymptomatically in the stem tissue until fruit ripening (Figure 1) [2,5–7]. Unripe fruits are resistant to SER [7]. This resistance is compromised when fruit ripening initiates during fruit storage. During ripening, fruits undergo dramatic biochemical and physiological changes including ethylene emission in climacteric fruit and other phytohormone changes, accumulation of soluble sugar, cell wall loosening, a decrease in phytoanticipin and phytoalexin levels, a decline in inducible plant defense mechanisms, and changes in ambient host pH [1].

The endophytic pathogens probably sense the changes during fruit ripening and respond to them by switching from an endophytic, asymptomatic lifestyle termed 'quiescent' or 'latent' stage, to an aggressive necrotrophic stage, causing SER [5–7]. These physiological alterations modify the endophytic microorganism's environment in the fruit and consequently influence the fruit's susceptibility to SER [6]. At the early ripening stage, SER symptoms appear as a small dark-brown to black spot at the fruit stem-end. In advanced stages of ripening, SER progresses to decay, resulting in fruit discoloration, brown flesh, and fruit softening [8,9]. A positive correlation was found between length of ripening time and severity of several postharvest diseases, including SER, in avocado fruit [10]. Indeed, fruits that ripen faster have less SER than fruits that are slower to ripen [11], and therefore the longer ripening time in avocado increases the time available for fungal colonization and the opportunity for SER symptoms to develop [12].

SER occurs in various fruits, but it has mainly been studied in mango and avocado. Interestingly, similar pathogens cause SER in both mango and avocado fruit. The major pathogens causing SER in mango are illustrated in Figure 2 and including mainly *Botryosphaeria*-related species such as: *Dothiorella dominicana*, *Dothiorella mangiferae*, *Lasiodiplodia theobromae*, *Neofusicoccum* spp., *Phomopsis mangiferae*, *Cytosphaera mangiferae* and *Pestalotiopsis* sp. [5,13], and *Alternaria alternata* as well as *Colletotrichum gloeosporioides* [2,5]. Similarly, SER-causing pathogens of avocado include: *Colletotrichum gloeosporioides*, *Alternaria alternata* and various species of the *Botryosphaeria* family as described for mango [14]. The genus *Lasiodiplodia* is an emerging pathogen, associated with SER worldwide. In recent years, there has been a rise in reports of this pathogen causing heavy losses to the fruit industry in Brazil [15], China [16], Peru [17], and India [18]. As *Lasiodiplodia* prefer higher temperatures and attack during plant stress, this rise might be connected with global climate warming and has to be further explored.

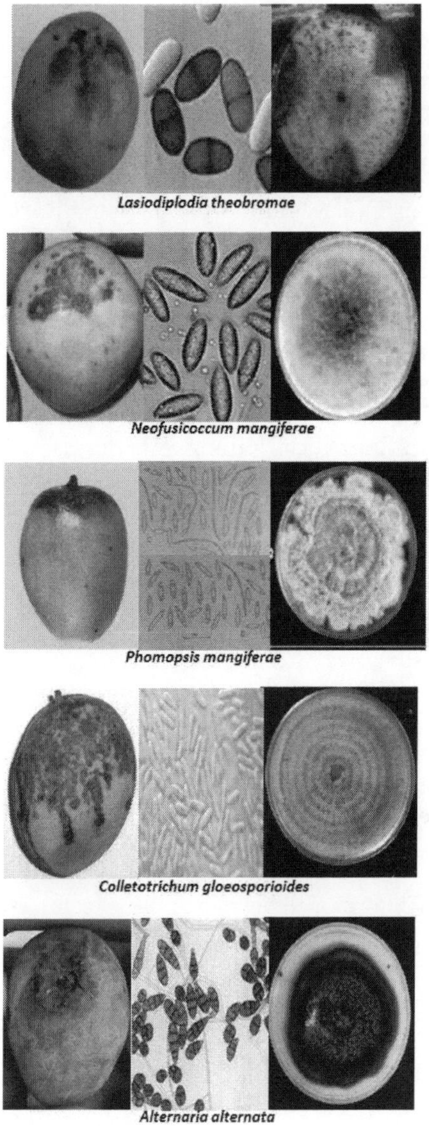

Figure 2. Characterization of fungal pathogens that cause stem-end rot in mango fruit. Left column: typical disease symptoms. Middle column: typical conidia. Right column: fungal growth on PDA media.

2. Endophytic Community in Fruit Stem-End

Not all fungi present in the stem-end make the transition from endophytic to necrotrophic lifestyle and become pathogenic during fruit ripening. The plant stem is populated with various species of microorganisms, including fungi, yeast and bacteria, most of which are not pathogenic [6,19]. These microorganisms can live in symbiosis or mutualism with the plant. They are termed 'endophytes' if they colonize the plant tissue internally. Endophytes persist in the plant tissue without causing any apparent symptoms or damage [20].

Diverse endophytic communities are considered to be important in maintaining a healthy plant biosystem. Little is known about the endophytic community of the fruit stem-end. However, endophytic microbiomes have been well studied in other plant organs, such as seeds, bark, foliage and roots [20,21]. Fungal endophytes are found in most plant families. Tropical trees represent hotspots of fungal species diversity, containing numerous species that have not been recovered [22]. The tree bark of Indian Bael trees was shown to have more endophytic fungi than its leaves or roots [23]. Thus, the fruit stem-end microbiome probably contains various microorganisms that should be further studied.

Recent advances in DNA sequencing and "omics" technologies have enabled evaluating the diversity and understanding the function of microbial communities existing within the plant tissue. In recent years, a few publications have also begun exploring the fruit stem-end microbiome in apple [19,24] and mango [6]. During storage and fruit ripening, the microbial community changes and the abundance of pathogenic fungi increases dramatically, along with increasing incidence of SER. These findings highlight the existence of an inherent mechanism by which pathogens have a quiescent endophytic stage and become active and cause disease when the composition of the microbiota changes in response to fruit ripening and storage [6].

3. Factors and Treatments Affecting SER

Postharvest disease management has the goal of preserving fruit quality without disease until consumption. Thus, management approaches are aimed at preventing, suppressing or delaying disease symptoms during storage [25]. Postharvest disease management in general, and SER in particular, can be achieved by several main approaches, such as chemical, biological and physical treatments that directly inhibit fungal pathogens on the one hand, or regulate fruit resistance on the other (Table 1).

3.1 Inhibition of Fruit Ripening

Ethylene is the main phytohormone controlling most of the events associated with the climacteric fruit-ripening process. Other phytohormones, such as auxin and abscisic acid, are also closely associated with fruit ripening [26]. Since there is a positive correlation between fruit ripening and postharvest decay, several studies have evaluated the potential application of phytohormones to prevent postharvest decay. Indeed, application of the auxin derivative 2,4-dichlorophenoxyacetic acid (2,4-D) reduced fruit ripening and prevented abscission of the stem-end in citrus and mango fruit, which reduced SER [27,28]. Other studies assessed the effect of postharvest application of ethylene receptor inhibitor, 1-MCP, which delays fruit senescence and prolongs storage [29,30], on the inhibition of postharvest decay. However, because ethylene plays a dual role, in fruit ripening and in the fruit defense response [1], those studies yielded conflicting findings. In Indian jujube fruit [31,32] and avocado fruit [33], 1-MCP treatment reduced fungal pathogen rot and SER. However, other studies in citrus [34], mango [35], and avocado [36] showed that 1-MCP promotes fruit susceptibility to SER pathogens. It seems that 1-MCP could affect fruits susceptibility in a concentration and timely dependent manner [37–39]. Thus, small amounts of ethylene are probably necessary to maintain fruit resistance to pathogens [1], and a high concentration of 1-MCP probably both delays the ripening process and hampers the fruit's natural defense.

3.2 Harvesting with Stem

One of the most intriguing ways to reduce SER is derived from a simple harvest practice, i.e., harvesting fruit with short stems (pedicel) using secateurs as opposed to the common practice of detaching the fruit, which leaves no stem at all. Surprisingly, this minor change in harvesting practice had a major and significant impact on reducing SER incidence in mango and avocado fruit [40]. Similarly, harvesting mango with long pedicels reduced SER in comparison to harvesting with short pedicels [41]. Interestingly, in the sap, there are some compounds with antimicrobial properties, which could cause the difference in SER incidence. For example, in mango sap, the alk(en)ylresorcinols (5-n-heptadecenylresorcinol and 5-n-pentadecylresorcinol) have antimicrobial and antifungal activities, especially against *Alternaria alternata* [42,43]. Furthermore, 'Kensington Pride' mango fruit stored with 2- to 3-cm long stems had significantly more resorcinol in their peel and smaller anthracnose lesions than de-sapped fruit [44]. Similarly, mango cultivar with higher sap flow had less incidence of anthracnose [45]. Therefore, it seems that when harvesting fruit with stems, more sap that contains antifungal compounds is left in the fruit stem and peel, leading to decreased postharvest side decay and SER.

3.3 Chemical Treatments

Fungicides are generally the most traditional and effective strategy for controlling postharvest diseases [46]. Fungicide type and timing of application depend on the target pathogen and its lifecycle. A variety of pre- and postharvest fungicidal treatments were suggested to reduce or delay the onset of SER.

3.3.2. Preharvest Chemical Control

Preharvest application of fungicide is efficient in reducing SER, while the fungicide residue decline with time. Preharvest sprays with Benlate applied to 'Hamlin' orange trees was found to eliminate SER in orange fruit harvested a week later, and efficiently reduced green mold (*Penicillium digitatum*) in fruit harvested six weeks after the spray application [47]. Preharvest sprays with various chemicals applied to 'Fuerte' avocado fruit were efficient in controlling SER and anthracnose [48]. Difolatan was found to be the most effective product closely followed by Cu-hydroxide and Baycor. Good control was also achieved with Aliette, Benlate and Cu-oxychloride [48].

Since SER-causing pathogens penetrate mainly during flowering and colonize the fruit stem before harvest [5], targeting the flowering stage during fungal penetration could reduce SER. Preharvest spray application of copper oxychloride, combined with mancozeb, from flowering until harvest, controls most mango postharvest diseases [49]. Diskin, Feygenberg, Maurer and Alkan [4] recently showed that fungicide application of Luna Tranquility (fluopyram and pyrimethanil) or Switch (fludioxonil and cyprodinil) during flowering, as *Lasiodiplodia* penetrates, significantly reduces the incidence and severity of postharvest SER and side decay in mango fruits. These results suggest that fungicide application during flowering reduced the penetration and initial colonization of pathogenic fungi in the fruit stem-end, which shifted the fruit stem-end microbiome toward a more diverse and less pathogenic community, leading to a reduction in SER incidence [4].

3.3.2. Postharvest Chemical Control

Postharvest fungicidal treatments are more common for controlling SER and can be applied by dipping or spraying, or in waxes or coatings. Prochloraz, a nonsystemic imidazole, is a well-recognized fungicide that is used commercially for controlling postharvest diseases in avocado and mango fruit [50–52]. However, application of prochloraz has been reported to be more effective against side decay (anthracnose), and less effective against SER. In general, benzimidazole fungicides, including benomyl and thiabendazole, have the advantage of also being effective against SER caused by *Lasiodiplodia theobromae* on mango, whereas imidazoles such as prochloraz and imazalil are not effective for SER control [53]. Similarly, Plan, et al. [54] found that benomyl is more effective than prochloraz and pyrimethanil for controlling mango SER caused by *Botryosphaeria*. Fludioxonil was more effective against mango fruit SER, whereas prochloraz was more effective against anthracnose [55]. The combination of prochloraz and fludioxonil was most effective at controlling both postharvest diseases—anthracnose and SER—in 'Kent' mango fruit [56]. In a comparative study, the efficacy of six fungicides—carbendazim, azoxystrobin, tebuconazole + trifloxystrobin, difenoconazole, thiabendazole and propiconazole—was assayed on mango artificially inoculated with *Lasiodiplodia theobromae* inoculum. They showed that carbendazim, followed by thiabendazole were highly effective at inhibiting the mycelial growth of *Lasiodiplodia theobromae*. They also reported that tebuconazole + trifloxystrobin, azoxystrobin and carbendazim significantly reduce SER disease severity on mango fruit [57].

A comparative study of azoxystrobin, fludioxonil, pyrimethanil, imazalil and thiabendazole against *Diplodia* SER in citrus fruit showed highest effectiveness for thiabendazole, imazalil and fludioxonil [58]. In artificially inoculated lemons, Phomopsis stem-end rot caused by *Diaporthe citri* was effectively controlled by low toxicity salts as potassium sorbate and potassium phosphite at 20°C, although Diplodia stem-end rot caused by *Lasiodiplodia theobromae* was partially controlled only by potassium sorbate [59]. Recently, the conventional fungicides imazalil and thiabendazole were found to be effective at controlling Diplodia SER caused by *Lasiodiplodia theobromae*. The best control of Diplodia SER was achieved by immersion in thiabendazole at pH 5 and 20°C. It was concluded that thiabendazole application for lemon treatment is the best alternative to controlling SER and should replace carbendazim, which is, however, not allowed in the European Union [60].

3.4 Biological Control

Postharvest applications of chemical fungicides are probably the best means of controlling postharvest decay. However, there is an increase in public concern over the use of chemical fungicides due to their negative effects on the environment and consumer health. In addition, repeated use of fungicide could lead to the development of resistant strains of pathogens. Therefore, there is a need for alternative approaches for postharvest disease management [61]. Biological control, which use microbial antagonists such as bacteria, yeast and fungi against postharvest pathogens, is an efficient strategy for controlling SER [62,63]. While most of the postharvest microbial antagonist research has focused on controlling side decay of fruit, some of those microbial antagonists have biological control activity against SER-causing pathogens. Timing of application is of crucial importance in biological control programs and significantly influence on control efficiency. Thus antagonists can be applied pre and post-harvest.

3.4.1. Preharvest Biological Control

Several preharvest treatments were studied in order to reduce SER disease. Preharvest application of *Bacillus subtilis* were found to be effective in controlling several postharvest decays as anthracnose, SER, and *Dothiorella–Colletotrichum* complex in avocado fruits [64]. To control SER disease using biological control strategy it could be important to apply the antagonist during inflorescence and flowering, when SER-causing pathogens penetrate. Interestingly, *B. subtilis* was found to attach and colonize avocado flowers and interfere with SER-causing pathogen penetration and initial colonization, by attach the conidia and hyphae of SER-causing pathogens and cause cell degradation [65]. Indeed, application of *B. subtilis* during mango flowering reduced pathogenic fungal colonization in the stem-end and reduced mango SER [4].

3.4.2. Postharvest Biological Control

Postharvest application of *Bacillus licheniformis* reduced mango anthracnose and SER [66]. In banana, anthracnose and crown rots caused by *Colletotrichum* is controlled by the biological agents *Burkholderia cepacia*, *Pichia anomala*, *Pseudomonas* sp. and *Candida oleophila* [67,68]. In addition, *Trichoderma harzianum* was found to reduce SER caused by *Lasiodiplodia theobromae* in Rambutan fruit [69], and *Trichoderma viride* was reported to control SER caused by *L. theobromae* in mango fruit [70]. Thus, a variety of microbial agents have been found effective at controlling fruit SER-causing pathogens. The antagonistic mechanism of the various microorganisms could include competition for nutrients, production of antibiotics, direct parasitism or induction of fruit resistance [63].

3.5 Plant Extracts

Another approach for SER management is plant extracts, which proved to be a safe alternative to control postharvest diseases and are listed as food additives by the U.S. FDA [71,72]. The efficiency of plant volatiles has been demonstrated for reducing postharvest disease incidence, including prolonged shelf life and improved fruit quality in mangoes by hexanal [73] and in citrus by citral [74,75].

Many studies have revealed the antifungal potential of plant extracts against a range of fungal pathogens [76–78]. For example, *Moringa oleifera*, *Syzygium aromaticum* and *Cinnamomum zeylanicum* showed significant antifungal activity against mycelial growth of fungal pathogens that cause SER, and a reduction in SER development in mango fruit [79]. Similarly, a comparative analysis of plant extracts showed that extracts of *Datura stramonium* and *Eucalyptus camaldulensis* efficiently reduce the radial growth of *Lasiodiplodia* isolates in vitro [80]. Another comparative study showed that thyme oil vapours, and clove and cinnamon oil completely inhibited *Colletotrichum gloeosporioides* and *Lasiodiplodia theobromae* growth in vitro. They also showed that thyme oil significantly inhibited the postharvest pathogens on mango fruits after storage at 25 °C for six days [81].

Plant extracts have also been reported as inducers of the defense response in fruit against potential pathogens [82,83]. Obianom and Sivakumar [84], recently showed that a combination of prochloraz and 0.1% (*v*/*v*) thyme oil significantly reduces anthracnose and SER in the 'Fuerte' avocado. They also reported that the combined treatment induces activity of defense enzymes in 'Fuerte' avocados inoculated with *Lasiodiplodia theobromae* and *Colletotrichum gloeosporioides*. In another study, dipping treatment with combination of bacterial antagonists, *Bacillus subtilis* and hexanal induced systemic resistance of mango fruits against *Lasiodiplodia theobromae* by inducing several defense-related enzymes in mango [85].

3.6 Physical Control

Physical technologies, including heat treatment, irradiation and cold-temperature storage, are also common and safe approaches for SER control. Cold storage is one of the best ways to delay fruit ripening and therefore decrease postharvest decay. The effects of temperature on endophytes have been poorly characterized [86]. However, each fungus has an optimal temperature and a temperature that limits their hyphal growth, conidial germination and pathogenicity [87]. On the other hand, each fruit has an optimal storage temperature. Storage below this temperature leads to chilling injuries, and storage above this temperature leads to faster ripening. Indeed, storage of the 'Hass' avocado, for example, at a temperature higher than 6 °C increased fruit ripening and the occurrence of SER [11], while storage at suboptimal temperature (lower than 5 °C) also increased the occurrence of SER [88].

Gamma irradiation can kill microorganisms by damaging their DNA [89], and can even be used to extend the shelf life of foods [90]. However, gamma irradiation did not reduce mango or citrus SER [91,92], or only mildly reduced SER [93], whereas a combination of hot water with gamma irradiation significantly reduced mango SER and anthracnose [92,93]. UV-C was found to control various fungal pathogens, including SER-causing pathogens [94,95], by inducing fruit resistance [96], and can also be considered for the organic market. Similarly, red mango fruit that was exposed to sunlight in the orchard accumulated anthocyanin and was more resistant to SER than fruit that developed within the tree canopy [6]. Therefore, pruning and exposure of fruit to sunlight could be a good method for reducing postharvest SER.

Heat treatment can induce the fruit's natural resistance, remove the unattached pathogens and cause spreading of the fruit's waxy covering, leading to a reduction in postharvest diseases, including SER, reduction in chilling injury and improvement of shelf life [97,98]. Different heat-treatment approaches include hot-water dipping and rinsing; hot vapor and dry-air treatments have been suggested to reduce postharvest diseases via induction of a defense response [99]. Hot-water immersion also reduced SER of papaya and mango [92,93,100,101], albeit with less efficiency than its effect on anthracnose. Therefore, various studies have integrated hot-water treatments in their postharvest treatment protocol.

Table 1. Summarize the technologies used for controlling fruit SER.

Treatment		SER-Causing Target Pathogen	Fruit Host	References
Preharvest chemical control	Benlate	*Lasiodiplodia theobromae*	Citrus	[47]
	Copper-based chemicals	*Lasiodiplodia* spp.	Mango	[49]
	Fluopyram + pyrimethanil or fludioxonil + cyprodinil	*Lasiodiplodia theobromae*	Mango	[4]
Postharvest chemical control	Benzimidazole: benomyl or thiabendazole	*Lasiodiplodia theobromae*	Mango	[53]
	Prochloraz and fludioxonil	*Lasiodiplodia theobromae*	Mango	[56]
	Tebuconazole + trifloxystrobin, azoxystrobin or carbendazim	*Lasiodiplodia theobromae*	Mango	[57]
	Thiabendazole, imazalil and fludioxonil	*Diplodia natalensis*	Mango	[58]
	Salts: potassium sorbate and potassium phosphite	*Diaporthe citri*, *Lasiodiplodia theobromae*	Lemon	[59]
	Thiabendazole	*Lasiodiplodia theobromae*	Lemon	[60]
Preharvest biological control	*Bacillus subtilis*	*Lasiodiplodia theobromae*	Avocado	[64,65]
	Bacillus subtilis	*Lasiodiplodia theobromae*	Mango	[4]
Postharvest biological control	*Bacillus licheniformis*	*Botryosphaeria* spp. and *L. theobromae*	Mango	[66]
	Trichoderma harzianum	*Lasiodiplodia theobromae*	Rambutan	[69]
	Trichoderma viride	*Lasiodiplodia theobromae*	Mango	[70]
Plant extracts	*Thyme oil* vapors	*Lasiodiplodia theobromae* and *Colletotrichum gloeosporioides*	Mango	[81]
	Combined prochloraz and *thyme oil*	*Lasiodiplodia theobromae* and *Colletotrichum gloeosporioides*	Avocado	[84]
	Moringa oleifera, *Syzygium aromaticum* and *Cinnamomum zeylanicum*	*Lasiodiplodia theobromae*, *Phomopsis mangiferae*, *Colletotrichum gloeosporioides*	Mango	[79]
Fruit Ripening inhibition	1-MCP	*Lasiodiplodia theobromae*	Jujube fruit	[31,32]
	1-MCP	*Lasiodiplodia theobromae*	Avocado	[33]
	2,4-D	*Phomopsis* spp. and *Lasiodiplodia* spp.	Citrus and mango	[27,28]
Agrotechnical methods	Harvesting with short stems (pedicel)	*Lasiodiplodia theobromae*	Mango	[40,41]
	Pruning, exposure to sunlight	*Lasiodiplodia theobromae*	Mango	[6]
Physical treatment	Hot-water or combined hot water and gamma irradiation	*Lasiodiplodia theobromae*	Papya and Mango	[92,93,101]

Table 1. *Cont.*

	Treatment	SER-Causing Target Pathogen	Fruit Host	References
Combined	Benomyl dip in hot water	*Dothiorella dominicana* and *Lasiodiplodia theobromae*	Mango	[102–104]
	Combined HWB along with prochloraz followed by 2,4-D	*Phomopsis* spp. and *Lasiodiplodia* spp.	Mango	[28]
	Hot water treatment with benomyl followed by a prochloraz	*Dothiorella dominicana*	Mango	[104]
	Combined *Bacillus subtilis* and hexanal	*Lasiodiplodia theobromae*	Mango	[85]

3.7 Combined Treatments

With the emergence of various fungicide-resistant isolates, and the specificity of each fungicide on the one hand, and difficulty achieving complete protection against postharvest decay using only physical or natural control on the other, one treatment alone cannot generally provide complete protection against all postharvest diseases. Thus, a combination of strategies must be applied to enhance the efficiency of coping with various postharvest diseases.

Integrating heat treatment with some chemical compounds resulted in a synergistic increase in control effectiveness leading to a significant decline in the chemical concentration needed to control postharvest decay. In several countries, mangoes were treated by immersing the fruits for 5 min at 52 °C combined with benomyl [76]. Indeed, benomyl dip in hot water was reported to efficiently control SER caused by both *Dothiorella dominicana* and *Lasiodiplodia theobromae* [102–104]. Carbendazim can also be applied with hot water (52 °C) for SER and anthracnose control [105]. Hot thiabendazole is generally effective at controlling SER, but provides poor control of anthracnose [102].

In mango, a combination of hot water brushing (HWB) along with prochloraz followed by 2,4-D significantly reduced SER and side decay by 50–70% and improved mango fruit quality during prolonged storage [28]. This combination reduced the incidence of SER from 86 to 10% in 'Tommy Atkins' mangoes. One of the common treatments for mangoes includes the combination of chlorine sterilization followed by HWB, then acidic prochloraz application followed by waxing [106]. Another combination offered in Australia is hot-water treatment with benomyl followed by a prochloraz spray, which provides effective control of anthracnose, SER and alternaria rot in mangoes [104].

Efficient control of postharvest disease, including SER, should therefore integrate several approaches. These can include preharvest chemical or biological application, harvesting with short stems, removing the sap, surface sterilization, hot-water treatment, chemical or biological fungicide treatment, waxing or coating, ripening inhibition, and cold storage. However, not all of these treatments are necessary if the disease rate is low or if the storage period is relatively short. In addition, each approach costs money. Thus, each packing house must customize their own protocols to control postharvest diseases and prolong fruit quality during storage.

4. Fruit Stem-End Microbiome and Modern Molecular Tools

The last decade was accompanied with major advantages in high-throughput sequencing of DNA and RNA and computational mapping. Those methods could be applied to fast sequencing of fungal genomes, fruit-pathogen interaction in the transcriptional level and study the dynamics of microorganism in the fruit stem-end. Today, the genomes of most of the hosts and SER causing pathogens are available, which will enable transcriptome analysis that could open new insights for better understanding the host effective defense response and the switch of pathogenic fungi from endophytic to pathogenic stage. This transcriptome analysis could lead to the development of new control methods.

The microbiome consists of all of the microorganisms (fungi and bacteria) inhabiting the plant tissue. The advances in high-throughput sequencing of DNA have been applied to study the plant microbiome. In apples, variety and rootstock modulate the endophytic microbiome, suggesting coevolution of a specific genotype with its microbiome [19]. When these methods were used to study the mango stem-end microbiome during postharvest storage and disease development, healthier stem-ends (with a lower incidence of SER) showed more diverse microbial communities [6]. In contrast, an increase in SER incidence was correlated with a reduction in microbiome diversity and an expansion of one or several fungal pathogen families, such as *Pleosporaceae* and *Botryosphaeriaceae*. In addition, the increase in fungal abundance and SER was correlated with an increase in the chitin-degrading *Chitinophagaceae* bacteria and a reduction in biocontrol agents [6]. This implies that the stem-end microbiota is a dynamic system that can be modified to control SER-causing pathogens.

5. Summary and Future Directions

Several fungal pathogens can cause SER. Their conidia or spores are carried by wind or water and penetrate through natural openings in the fruit. The penetrated fungi endophytically colonize the phloem or xylem of the pedicel and localize in the tissue connecting the stem-end to the fruit body (Figure 1). During ripening, the fruit becomes susceptible to various fungal pathogens (Figure 2) that switch from endophytic stage to necrotrophic stage and cause SER (Figure 1). Although SER disease leads to significant fruit loss, the basic science of fruit SER is largely unknown. Nevertheless, studies have shown that different treatments can decrease the incidence of SER by directly inhibiting the fungal growth or indirectly inducing host resistance (Table 1), or by indirectly changing the stem-end microbiome to a more diverse and less pathogenic community. With the increased availability of new tools such as deep sequencing, new studies are expected to emerge, leading to a better understanding of the host–pathogen interaction and the stem-end holobiont, which could lead to the development of new means to reduce fruit SER.

References

1. Alkan, N.; Fortes, A.M. Insights into molecular and metabolic events associated with fruit response to post-harvest fungal pathogens. *Front. Plant Sci.* **2015**, *6*, 889. [CrossRef] [PubMed]

2. Prusky, D.; Kobiler, I.; Miyara, I.; Alkan, N. Fruit diseasess. In *The Mango, Botany, Production and Uses*, 2nd ed.; Litz, R., Ed.; CABI International: Wallingford, UK, 2009; pp. 210–231.

3. Alkan, N.; Kumar, P. Postharvest storage management of mango fruit. In *Achieving Sustainable Cultivation of Mango*; Galán Saúco, V., Lu, P., Eds.; Burleigh Dodds Science Publishing: Cambridge, UK, 2018; pp. 377–402.

4. Diskin, S.; Feygenberg, O.; Maurer, D.; Alkan, N. Biological and chemical application during flowering increase fruit count, yield and reduce postharvest decay in mango fruit. *Front. Plant Sci.* submitted.

5. Johnson, G.I.; Mead, A.J.; Cooke, A.W.; Dean, J.R. Mango stem end rot pathogens—Fruit infection by endophytic colonisation of the inflorescence and pedicel. *Ann. Appl. Biol.* **1992**, *120*, 225–234. [CrossRef]

6. Diskin, S.; Feygenberg, O.; Maurer, D.; Droby, S.; Prusky, D.; Alkan, N. Microbiome alterations are correlated with occurrence of postharvest stem-end rot in mango fruit. *Phytobiomes* **2017**, *1*, 117–127. [CrossRef]

7. Prusky, D.; Alkan, N.; Mengiste, T.; Fluhr, R. Quiescent and necrotrophic lifestyle choice during postharvest disease development. *Annu. Rev. Phytopathol.* **2013**, *51*, 155–176. [CrossRef] [PubMed]

8. Ploetz, R.C. *Compendium of Tropical Fruit Diseases*; APS Press: St. Paul, MN, USA, 1994.

9. Pérez-Jiménez, R.M. Significant avocado diseases caused by fungi and oomycetes. *Eur. J. Plant Sci. Biotechnol.* **2008**, *2*, 1–24.

10. Darvas, J.; Kotze, J.; Wehner, F. Effect of treatment after picking on the incidence of post-harvest fruit diseases of avocado. *Phytophylactica* **1990**, *22*, 93–96.

11. Hopkirk, G.; White, A.; Beever, D.; Forbes, S. Influence of postharvest temperatures and the rate of fruit ripening on internal postharvest rots and disorders of New Zealand 'Hass' avocado fruit. *N. Z. J. Crop Hortic. Sci.* **1994**, *22*, 305–311. [CrossRef]

12. Arpaia, M.L.; Collin, S.; Sievert, J.; Obenland, D. Influence of cold storage prior to and after ripening on quality factors and sensory attributes of 'Hass' avocados. *Postharvest Biol. Technol.* **2015**, *110*, 149–157. [CrossRef]

13. Johnson, G.I.; Mead, A.J.; Cooke, A.W.; Dean, J.R. Mango stem end rot pathogens—Infection levels between flowering and harvest. *Ann. Appl. Biol.* **1991**, *119*, 465–473. [CrossRef]

14. Hartill, W.; Everett, K. Inoculum sources and infection pathways of pathogens causing stem-end rots of 'Hass' avocado (*Persea americana*). *N. Z. J. Crop Hortic. Sci.* **2002**, *30*, 249–260. [CrossRef]

15. Coutinho, I.; Freire, F.; Lima, C.; Lima, J.; Gonçalves, F.; Machado, A.; Silva, A.; Cardoso, J. Diversity of genus *Lasiodiplodia* associated with perennial tropical fruit plants in northeastern Brazil. *Plant Pathol.* **2017**, *66*, 90–104. [CrossRef]

16. Zhang, H.; Wei, Y.; Qi, Y.; Pu, J.; Liu, X. First report of *Lasiodiplodia* Brasiliense associated with stem-end rot of mango in China. *Plant Dis.* **2018**, *102*, 679. [CrossRef]

17. Rodríguez-Gálvez, E.; Guerrero, P.; Barradas, C.; Crous, P.W.; Alves, A. Phylogeny and pathogenicity of *Lasiodiplodia* species associated with dieback of mango in Peru. *Fungal Biol.* **2017**, *121*, 452–465. [CrossRef] [PubMed]

18. Sathya, K.; Parthasarathy, S.; Thiribhuvanamala, G.; Prabakar, K. Morphological and molecular variability of *Lasiodiplodia theobromae* causing stem end rot of mango in Tamil Nadu, India. *Int. J. Pure Appl. Biosci.* **2017**, *5*, 1024–1031. [CrossRef]

19. Liu, J.; Abdelfattah, A.; Norelli, J.; Burchard, E.; Schena, L.; Droby, S.; Wisniewski, M. Apple endophytic microbiota of different rootstock/scion

combinations suggests a genotype-specific influence. *Microbiome* **2018**, *6*, 18. [CrossRef] [PubMed]

20. Wilson, D. Endophyte: The evolution of a term, and clarification of its use and definition. *Oikos* **1995**, *73*, 274–276. [CrossRef]
21. Porras-Alfaro, A.; Bayman, P. Hidden fungi, emergent properties: Endophytes and microbiomes. *Annu. Rev. Phytopathol.* **2011**, *49*, 291–315. [CrossRef] [PubMed]
22. Arnold, A.E. Understanding the diversity of foliar endophytic fungi: Progress, challenges, and frontiers. *Fungal Biol. Rev.* **2007**, *21*, 51–66. [CrossRef]
23. Gond, S.; Verma, V.; Kumar, A.; Kumar, V.; Kharwar, R. Study of endophytic fungal community from different parts of *Aegle marmelos* Correae (Rutaceae) from Varanasi (India). *World J. Microbiol. Biotechnol.* **2007**, *23*, 1371–1375. [CrossRef]
24. Abarenkov, K.; Henrik Nilsson, R.; Larsson, K.H.; Alexander, I.J.; Eberhardt, U.; Erland, S.; Høiland, K.; Kjøller, R.; Larsson, E.; Pennanen, T. The unite database for molecular identification of fungi–recent updates and future perspectives. *New Phytol.* **2010**, *186*, 281–285. [CrossRef] [PubMed]
25. Sommer, N.F. Role of controlled environments in suppression of postharvest diseases. *Can. J. Plant Pathol.* **1985**, *7*, 331–339. [CrossRef]
26. Kumar, R.; Khurana, A.; Sharma, A.K. Role of plant hormones and their interplay in development and ripening of fleshy fruits. *J. Exp. Bot.* **2014**, *65*, 4561–4575. [CrossRef] [PubMed]
27. Coggins, C.W., Jr. Use of growth regulators to delay maturity and prolong shelf life of citrus. In Proceedings of the Symposium on Growth Regulators in Fruit Production, St. Paul, MN, USA, 11–15 November 1972; pp. 469–472.
28. Kobiler, I.; Shalom, Y.; Roth, I.; Akerman, M.; Vinokur, Y.; Fuchs, Y.; Prusky, D. Effect of 2,4- dichlorophenoxyacetic acid on the incidence of side and stem end rots in mango fruits. *Postharvest Biol. Technol.* **2001**, *23*, 23–32. [CrossRef]
29. Watkins, C.B.; Nock, J.F. Effects of delayed controlled atmosphere storage of apples after rapid 1-MCP treatment. *HortScience* **2008**, *43*, 1087.
30. Watkins, C.B. The use of 1-methylcyclopropene (1-MCP) on fruits and vegetables. *Biotechnol. Adv.* **2006**, *24*, 389–409. [CrossRef] [PubMed]
31. Zhang, Z.; Tian, S.; Zhu, Z.; Xu, Y.; Qin, G. Effects of 1-methylcyclopropene (1-MCP) on ripening and resistance of jujube (*Zizyphus jujuba* cv. Huping) fruit against postharvest disease. *LWT-Food Sci. Technol.* **2012**, *45*, 13–19. [CrossRef]
32. Zhong, Q.; Xia, W. Effect of 1-methylcyclopropene and/or chitosan coating treatments on storage life and quality maintenance of Indian jujube fruit. *LWT-Food Sci. Technol.* **2007**, *40*, 404–411.
33. Diskin, S.; Feygenberg, O.; Maurer, D.; Prusky, D.; Droby, S.; Alkan, N. Postharvest treatments for stem-end rot in fruits. *Alon Hanotea* **2015**, *70*, 36–38.

34. Porat, R.; Weiss, B.; Cohen, L.; Daus, A.; Goren, R.; Droby, S. Effects of ethylene and 1-methylcyclopropene on the postharvest qualities of 'Shamouti' oranges. *Postharvest Biol. Technol.* **1999**, *15*, 155–163. [CrossRef]

35. Hofman, P.J.; Jobin-Decor, M.; Meiburg, G.F.; Macnish, A.J.; Joyce, D.C. Ripening and quality responses of avocado, custard apple, mango and papaya fruit to 1-methylcyclopropene. *Aust. J. Exp. Agric.* **2001**, *41*, 567–572. [CrossRef]

36. Woolf, A.B.; Requejo-Tapia, C.; Cox, K.A.; Jackman, R.C.; Gunson, A.; Arpaia, M.L.; White, A. 1-MCP reduces physiological storage disorders of 'Hass' avocados. *Postharvest Biol. Technol.* **2005**, *35*, 43–60. [CrossRef]

37. Dou, H.; Jones, S.; Ritenour, M. Influence of 1-MCP application and concentration on post-harvest peel disorders and incidence of decay in citrus fruit. *J. Hortic. Sci. Biotechnol.* **2005**, *80*, 786–792. [CrossRef]

38. Ku, V.V.V.; Wills, R.B.H.; Ben-Yehoshua, S. 1-methylcyclopropene can differentially affect the postharvest life of strawberries exposed to ethylene. *HortScience* **1999**, *34*, 119–120.

39. Blanco-Ulate, B.; Vincenti, E.; Powell, A.L.T.; Cantu, D. Tomato transcriptome and mutant analyses suggest a role for plant stress hormones in the interaction between fruit and *Botrytis cinerea. Front. Plant Sci.* **2013**, *4*, 142. [CrossRef] [PubMed]

40. Diskin, S.; Lurie, S.; Feygenberg, O.; Maurer, D.; Droby, S.; Prusky, D.; Alkan, N. Postharvest Microbiota Dynamics of Mango Fruit Stem-End. In Proceedings of the IV International Symposium on Postharvest Pathology, Skukuza, South Africa, 28 May–3 June 2017.

41. Sangchote, S. Botryodiplodia Stem End Rot of Mango and Its Control. In Proceedings of the III International Mango Symposium, Darwin, Australia, 25–29 September 1989; pp. 296–303.

42. Droby, S.; Prusky, D.; Jacoby, B.; Goldman, A. Induction of antifungal resorcinols in flesh of unripe mango fruits and its relation to latent infection by *Alternaria alternata. Physiol. Mol. Plant Pathol.* **1987**, *30*, 285–292. [CrossRef]

43. Negi, P.; John, S.K.; Rao, P.U. Antimicrobial activity of mango sap. *Eur. Food Res. Technol.* **2002**, *214*, 327–330.

44. Hassan, M.; Dann, E.; Irving, D.; Coates, L. Concentrations of constitutive alk(en) ylresorcinols in peel of commercial mango varieties and resistance to postharvest anthracnose. *Physiol. Mol. Plant Pathol.* **2007**, *71*, 158–165. [CrossRef]

45. Krishnapillai, N.; Wijeratnam, W. Sap volatile components in relation to susceptibility of anthracnose and aspergillus rot of mangoes (*Mangifera indica* L.). *J. Hortic. Sci. Biotechnol.* **2017**, *92*, 206–213. [CrossRef]

46. Singh, D.; Sharma, R. Postharvest diseases of fruit and vegetables and their management. In *Sustainable Pest Management*; Prasad, D., Ed.; Daya Publishing House: New Delhi, India, 2007; pp. 67–69.

47. Brown, G.E.; McCornack, A. Benlate, an experimental preharvest fungicide for control of postharvest citrus fruit decay. *Proc. Fla. State Hortic. Soc. 1969* **1970**, *82*, 39–43.

48. Darvas, J. Pre-harvest chemical control of post-harvest avocado diseases. *S. Afr. Avocado Growers' Assoc. Yearb.* **1981**, *7*, 57–59.

49. Lonsdale, J.; Kotze, J. Chemical control of mango blossom diseases and the effect on fruit set and yield. *Plant Dis. (USA)* **1993**, *77*, 558. [CrossRef]

50. Dodd, J.C.; Prusky, D.; Jeffries, P. Fruit Diseases. In *The Mango: Botany, Production and Uses*; CAB International: Wallingford, UK, 1997; pp. 257–280.

51. Muirhead, I.; Fitzell, R.; Davis, R.; Peterson, R. Post-harvest control of anthracnose and stem-end rots of fuerte avocados with prochloraz and other fungicides. *Aust. J. Exp. Agric.* **1982**, *22*, 441–446. [CrossRef]

52. Everett, K.; Owen, S.; Cutting, J. Testing efficacy of fungicides against postharvest pathogens of avocado (*Persea americana* cv. Hass). *N. Z. Plant Protect.* **2005**, *58*, 89–95.

53. Estrada, A.; Jeffries, P.; Dodd, J.C. Field evaluation of a predictive model to control anthracnose disease of mango in the Philippines. *Plant Pathol.* **1996**, *45*, 294–301. [CrossRef]

54. Plan, M.; Joyce, D.; Ogle, H.; Johnson, G. Mango stem-end rot (*Botryosphaeria dothidea*) disease control by partial-pressure infiltration of fungicides. *Aust. J. Exp. Agric.* **2002**, *42*, 625–629. [CrossRef]

55. Swart, S.; Serfontein, J.; Kalinnowski, J. Chemical control of postharvest diseases of mango-the effect of prochloraz, thiobendazole and fludioxonil on soft brown rot, stem-end rot and anthracnose. *S. Afr. Mango Growers' Assoc. Yearb.* **2002**, *22*, 55–62.

56. Swart, S.; Serfontein, J.; Swart, G.; Labuschagne, C. Chemical Control of Post-Harvest Diseases of Mango: The Effect of Fludioxonil and Prochloraz on Soft Brown Rot, Stem-End Rot and Anthracnose. In Proceedings of the VIII International Mango Symposium, Sun City, South Africa, 5–10 February 2006; pp. 503–510.

57. Syed, R.N.; Mansha, N.; Khaskheli, M.A.; Khanzada, M.A.; Lodhi, A.M. Chemical control of stem end rot of mango caused by *Lasiodiplodia theobromae*. *Pak. J. Phytopathol.* **2014**, *26*, 201–206.

58. Zhang, J. Comparative effects of postharvest fungicides for diplodia stem-end rot control of Florida citrus. *Proc. Fla. State Hortic. Soc.* **2012**, *125*, 243–247.

59. Cerioni, L.; Sepulveda, M.; Rubio-Ames, Z.; Volentini, S.I.; Rodríguez-Montelongo, L.; Smilanick, J.; Ramallo, J.; Rapisarda, V.A. Control of lemon postharvest diseases by low-toxicity salts combined with hydrogen peroxide and heat. *Postharvest Biol. Technol.* **2013**, *83*, 17–21. [CrossRef]

60. Cerioni, L.; Bennasar, P.B.; Lazarte, D.; Sepulveda, M.; Smilanick, J.L.; Ramallo, J.; Rapisarda, V.A. Conventional and reduced-risk fungicides to control

postharvest *Diplodia* and *Phomopsis* stem-end rot on lemons. *Sci. Hortic.* **2017**, *225*, 783–787. [CrossRef]

61. Wisniewski, M.; Droby, S.; Norelli, J.; Liu, J.; Schena, L. Alternative management technologies for postharvest disease control: The journey from simplicity to complexity. *Postharvest Biol. Technol.* **2016**, *122*, 3–10. [CrossRef]

62. Droby, S.; Wisniewski, M.; Teixidó, N.; Spadaro, D.; Jijakli, M.H. The science, development, and commercialization of postharvest biocontrol products. *Postharvest Biol. Technol.* **2016**, *122*, 22–29. [CrossRef]

63. Sharma, R.R.; Singh, D.; Singh, R. Biological control of postharvest diseases of fruits and vegetables by microbial antagonists: A review. *Biol. Control* **2009**, *50*, 205–221. [CrossRef]

64. Korsten, L.; De Villiers, E.; Wehner, F.; Kotzé, J. Field sprays of *Bacillus subtilis* and fungicides for control of preharvest fruit diseases of avocado in South Africa. *Plant Dis.* **1997**, *81*, 455–459. [CrossRef]

65. Demoz, B.T.; Korsten, L. Bacillus subtilis attachment, colonization, and survival on avocado flowers and its mode of action on stem-end rot pathogens. *Biol. Control* **2006**, *37*, 68–74. [CrossRef]

66. Govender, V.; Korsten, L.; Sivakumar, D. Semi-commercial evaluation of bacillus licheniformis to control mango postharvest diseases in South Africa. *Postharvest Biol. Technol.* **2005**, *38*, 57–65. [CrossRef]

67. De Costa, D.; Erabadupitiya, H. An integrated method to control postharvest diseases of banana using a member of the *Burkholderia cepacia* complex. *Postharvest Biol. Technol.* **2005**, *36*, 31–39. [CrossRef]

68. Lassois, L.; de Bellaire, L.D.L.; Jijakli, M.H. Biological control of crown rot of bananas with *Pichia anomala* strain K and *Candida oleophila* strain O. *Biol. Control* **2008**, *45*, 410–418. [CrossRef]

69. Sivakumar, D.; Wijeratnam, R.W.; Wijesundera, R.; Marikar, F.; Abeyesekere, M. Antagonistic effect of *Trichoderma harzianum* on postharvest pathogens of rambutan (*Nephelium lappaceum*). *Phytoparasitica* **2000**, *28*, 240. [CrossRef]

70. Kota, V.; Kulkarni, S.; Hegde, Y. Postharvest diseases of mango and their biological management. *J. Plant Dis. Sci.* **2006**, *1*, 186–188.

71. Abbaszadeh, S.; Sharifzadeh, A.; Shokri, H.; Khosravi, A.; Abbaszadeh, A. Antifungal efficacy of thymol, carvacrol, eugenol and menthol as alternative agents to control the growth of food-relevant fungi. *J. Mycol. Méd./J. Med. Mycol.* **2014**, *24*, e51–e56. [CrossRef] [PubMed]

72. Zhou, H.; Tao, N.; Jia, L. Antifungal activity of citral, octanal and α-terpineol against *Geotrichum citri-aurantii*. *Food Control* **2014**, *37*, 277–283. [CrossRef]

73. Anusuya, P.; Nagaraj, R.; Janavi, G.J.; Subramanian, K.S.; Paliyath, G.; Subramanian, J. Pre-harvest sprays of hexanal formulation for extending retention and shelf-life of mango (*Mangifera indica* L.) fruits. *Sci. Hortic.* **2016**, *211*, 231–240. [CrossRef]

74. Wuryatmo, E.; Able, A.; Ford, C.; Scott, E. Effect of volatile citral on the development of blue mould, green mould and sour rot on navel orange. *Aust. Plant Pathol.* **2014**, *43*, 403–411. [CrossRef]

75. Fan, F.; Tao, N.; Jia, L.; He, X. Use of citral incorporated in postharvest wax of citrus fruit as a botanical fungicide against *Penicillium digitatum*. *Postharvest Biol. Technol.* **2014**, *90*, 52–55. [CrossRef]

76. Sepiah, M. Effectiveness of hot water, hot benomyl and cooling on postharvest diseases of mango cv. Harumanis. *ASEAN Food J.* **1986**, *2*, 117–120.

77. Patni, C.; Kolte, S. Effect of some botanicals in management of Alternaria blight of rapeseed-mustard. *Ann. Plant Protect. Sci.* **2006**, *14*, 151–156.

78. Bhalodia, N.R.; Shukla, V. Antibacterial and antifungal activities from leaf extracts of *Cassia fistula* L.: An ethnomedicinal plant. *J. Adv. Pharm. Technol. Res.* **2011**, *2*, 104–109. [CrossRef] [PubMed]

79. Alam, M.W.; Rehman, A.; Sahi, S.T.; Malik, A.U. Exploitation of natural products as an alternative strategy to control stem end rot disease of mango fruit in Pakistan. *Pak. J. Agric. Sci.* **2017**, *54*, 503–511.

80. Ullah, S.F.; Hussain, Y.; Iram, S. Pathogenic characterization of *Lasiodiplodia* causing stem end rot of mango and its control using botanics. *Pak. J. Bot.* **2017**, *49*, 1605–1613.

81. Perumal, A.B.; Sellamuthu, P.S.; Nambiar, R.B.; Sadiku, E.R. Antifungal activity of five different essential oils in vapour phase for the control of *Colletotrichum gloeosporioides* and *Lasiodiplodia theobromae* in vitro and on mango. *Int. J. Food Sci. Technol.* **2016**, *51*, 411–418. [CrossRef]

82. Shao, X.; Cao, B.; Xu, F.; Xie, S.; Yu, D.; Wang, H. Effect of postharvest application of chitosan combined with clove oil against citrus green mold. *Postharvest Biol. Technol.* **2015**, *99*, 37–43. [CrossRef]

83. Wu, Y.; Duan, X.; Jing, G.; Ouyang, Q.; Tao, N. Cinnamaldehyde inhibits the mycelial growth of geotrichum citri-aurantii and induces defense responses against sour rot in citrus fruit. *Postharvest Biol. Technol.* **2017**, *129*, 23–28. [CrossRef]

84. Obianom, C.; Sivakumar, D. Differential response to combined prochloraz and thyme oil drench treatment in avocados against the control of anthracnose and stem-end rot. *Phytoparasitica* **2018**, *46*, 273–281. [CrossRef]

85. Seethapathy, P.; Gurudevan, T.; Subramanian, K.S.; Kuppusamy, P. Bacterial antagonists and hexanal-induced systemic resistance of mango fruits against *Lasiodiplodia theobromae* causing stem-end rot. *J. Plant Interact.* **2016**, *11*, 158–166. [CrossRef]

86. Stone, J.K.; Polishook, J.D.; White, J.F. Endophytic fungi. In *Biodiversity of Fungi*; Elsevier Academic Press: Burlington, MA, USA, 2004; pp. 241–270.

87. Ayerst, G. The effects of moisture and temperature on growth and spore germination in some fungi. *J. Stored Prod. Res.* **1969**, *5*, 127–141. [CrossRef]

88. Pesis, E.; Ackerman, M.; Ben-Arie, R.; Feygenberg, O.; Feng, X.; Apelbaum, A.; Goren, R.; Prusky, D. Ethylene involvement in chilling injury symptoms of avocado during cold storage. *Postharvest Biol. Technol.* **2002**, *24*, 171–181. [CrossRef]
89. Imlay, J.A.; Linn, S. DNA damage and oxygen radical toxicity. *Science* **1988**, *240*, 1302–1309. [CrossRef] [PubMed]
90. Farkas, J. Irradiation for better foods. *Trends Food Sci. Technol.* **2006**, *17*, 148–152. [CrossRef]
91. Ladaniya, M.; Singh, S.; Wadhawan, A. Response of 'nagpur' mandarin, 'mosambi' sweet orange and 'kagzi' acid lime to gamma radiation. *Radiat. Phys. Chem.* **2003**, *67*, 665–675. [CrossRef]
92. Spalding, D.; Reeder, W. Decay and acceptability of mangos treated with combinations of hot water, imazalil, and gamma radiation. *Plant Dis.* **1986**, *70*, 1149–1151. [CrossRef]
93. Rashid, M.; Grout, B.W.W.; Continella, A.; Mahmud, T. Low-dose gamma irradiation following hot water immersion of papaya (*Carica papaya* Linn.) fruits provides additional control of postharvest fungal infection to extend shelf life. *Radiat. Phys. Chem.* **2015**, *110*, 77–81. [CrossRef]
94. Stevens, C.; Wilson, C.; Lu, J.; Khan, V.; Chalutz, E.; Droby, S.; Kabwe, M.; Haung, Z.; Adeyeye, O.; Pusey, L. Plant hormesis induced by ultraviolet light-c for controlling postharvest diseases of tree fruits. *Crop Protect.* **1996**, *15*, 129–134. [CrossRef]
95. Stevens, C.; Khan, V.; Wilson, C.; Lu, J.; Chalutz, E.; Droby, S. The effect of fruit orientation of postharvest commodities following low dose ultraviolet light-c treatment on host induced resistance to decay. *Crop Protect.* **2005**, *24*, 756–759. [CrossRef]
96. Romanazzi, G.; Sanzani, S.M.; Bi, Y.; Tian, S.; Martínez, P.G.; Alkan, N. Induced resistance to control postharvest decay of fruit and vegetables. *Postharvest Biol. Technol.* **2016**, *122*, 82–94. [CrossRef]
97. Fallik, E. Prestorage hot water treatments (immersion, rinsing and brushing). *Postharvest Biol. Technol.* **2004**, *32*, 125–134. [CrossRef]
98. Lurie, S. Postharvest heat treatments. *Postharvest Biol. Technol.* **1998**, *14*, 257–269. [CrossRef]
99. Schirra, M.; D'hallewin, G.; Ben-Yehoshua, S.; Fallik, E. Host–pathogen interactions modulated by heat treatment. *Postharvest Biol. Technol.* **2000**, *21*, 71–85. [CrossRef]
100. Couey, H.; Alvarez, A.; Nelson, M. Comparison of hot-water spray and immersion treatments for control of postharvest decay of papaya [stem-end rots and anthracnose]. *Plant Dis.* **1984**, *68*, 436–437. [CrossRef]

101. Alvindia, D.G.; Acda, M.A. Revisiting the efficacy of hot water treatment in managing anthracnose and stem-end rot diseases of mango cv. 'Carabao'. *Crop Protect.* **2015**, *67*, 96–101. [CrossRef]

102. Coates, L.; Johnson, G.; Cooke, A. Postharvest disease control in mangoes using high humidity hot air and fungicide treatments. *Ann. Appl. Biol.* **1993**, *123*, 441–448. [CrossRef]

103. Johnson, G.; Muirhead, I.; Rappel, L. Mango post-harvest disease control: A review of research in Australia, Malaysia and Thailand. *ASEAN Food J.* **1989**, *4*, 139–141.

104. Johnson, G.; Sangchote, S.; Cooke, A. Control of stem end rot (*Dothiorella dominicana*) and other postharvest diseases of mangoes (cv. Kensington pride) during short-and long-term storage. *Trop. Agric. Trinidad Tobago* **1990**, *67*, 183–187.

105. Johnson, G.; Hofman, P. *Postharvest Technology and Quarantine Treatments*; CAB International: Wallingford, UK, 2009.

106. Prusky, D.; Fuchs, Y.; Kobiler, I.; Roth, I.; Weksler, A.; Shalom, Y.; Fallik, E.; Zauberman, G.; Pesis, E.; Akerman, M. Effect of hot water brushing, prochloraz treatment and waxing on the incidence of black spot decay caused by *Alternaria alternata* in mango fruits. *Postharvest Biol. Technol.* **1999**, *15*, 165–174. [CrossRef]

Chapter-9

Postharvest Treatments with GRAS Salts to Control Fresh Fruit Decay

1. Introduction

Pathogenic filamentous fungi are the most important causal agents of postharvest decay of fresh fruits. Depending on the fruit species, cultivar, and a wide range of pre- and postharvest factors and conditions, the incidence of fungal decay can cause considerable economic losses to growers and traders, especially if the produce is intended for export markets. Wholesale buyers often reject fruit loads if decay is found in export shipments, and furthermore, they may charge the producer for the transport and handling costs [1].

Every fresh fruit commodity is prone to decay caused by pathogenic fungi. Depending on the origin and characteristics of the infection, fungi that cause postharvest diseases can be classified into two general groups: (i) those that infect the fruit in the field and remain latent until their development after harvest, and (ii) those that infect the fruit through rind microwounds or injuries inflicted during harvest, transportation, postharvest handling, and commercialization [2]. Typically, fungal species that cause latent infections can also cause wound infections near or after harvest in the same fruit commodity under certain conditions, but the opposite is not true, and many economically important wound pathogens can only infect fruit if its peel is broken, therefore they are known as strict wound pathogens. This is the case of *Penicillium* spp., the causal agents of blue or green molds on many relevant fruit commodities.

For example, blue mold, caused by the species *Penicillium expansum* L., can lead to significant postharvest losses of apple, pear, stone fruits, kiwifruit, many berries, pomegranate, persimmon, and other subtropical and tropical fruits [3–5]. The two postharvest diseases that account for the largest decay losses of citrus fruits worldwide are green and blue molds, caused by *Penicillium digitatum* (Pers.:Fr.) Sacc. and *Penicillium italicum* Wehmer, respectively [6]. It is clear from these denominations that the common names of postharvest diseases are based on the symptoms they produce. Other important wound pathogens are *Geotrichum* spp., which cause sour rot on different horticultural products, and *Rhizopus* spp., which are dangerous postharvest pathogens that cause soft rot or Rhizopus rot on a wide range of fruit hosts. The most important species are *Geotrichum candidum* L., which attacks stone fruits such as peach, nectarine, plum, and sweet cherry, but also tomato and other fruit-like vegetables; and *Geotrichum citri-aurantii* (Ferraris) Butler, the cause of sour rot of citrus fruits. *Rhizopus stolonifer* (Ehrenb.) Vuill. is the most important species causing fruit soft rot [7–9].

Among latent pathogens, *Botrytis cinerea* Pers.:Fr., which causes gray mold, is the main causal agent of postharvest decay of grape, strawberry, blueberry and other small berries, kiwifruit, pomegranate, fig, etc., but also attacks many other fruits and vegetables [10,11]. *Colletotrichum* spp. cause anthracnose on many subtropical and tropical fruits, such as citrus fruit, avocado, persimmon, banana, papaya, mango, and pineapple. The most important species are *C. gloeosporioides* (Penz.) Penz. & Sacc., *C. acutatum* J.H. Simmonds, and *C. horii* B.S. Weir & P.R. Johnst. [12,13]. *Alternaria* spp., which causes black spot, and *Lasiodiplodia theobromae* (Pat.) Griffon and Maubl. and other Botryosphaeriaceae pathogens, which cause stem-end rot, are also economically relevant for the worldwide industry of subtropical and tropical fruit crops [14,15]. The species *Alternaria alternata* (Fr.) Keissl. is also an important postharvest pathogen of pome fruits, persimmon, loquat, and pomegranate, among many other fruits [16]. Species of *Monilinia*, especially *M. fructicola* (Winter) Honey, *M. laxa* (Aderhold & Ruhland) Honey, and *M. fructigena* (Aderhold & Ruhland) Honey, cause field and postharvest brown rot of stone fruits and are generally considered among the most substantial limiting factors of the yield of these fruit crops [17]. *Monilinia* spp. can also cause postharvest disease on pome fruits [18].

For many years now, fruit postharvest diseases and the economic losses they cause have been contained worldwide through postharvest applications of synthetic chemical fungicides in commercial packinghouses. Active ingredients such as imazalil (IMZ), pyrimethanil (PYR), fludioxonil (FLU), thiabendazole (TBZ), and others are being extensively used as cost-effective means of postharvest decay control in conventional horticulture. However, the massive and continuous use of these chemicals is increasingly leading to significant problems, such as human health issues and environmental contamination due to chemical residues, reduced efficacy of many synthetic fungicides due to the proliferation of resistant fungal biotypes, and restricted access to new high-value organic markets or traditional export markets that are now demanding products with lower levels of pesticides to satisfy consumer demands. Therefore, consumer trends and legislative updates clearly favor a reduction in the use of conventional fungicides, which makes it necessary to potentiate research to develop and implement alternative approaches and novel technologies for the control of postharvest diseases. If conventional chemicals are not available, effective control will need to adopt integrated strategies in which, besides new nonpolluting postharvest antifungal treatments, all factors affecting disease epidemiology and incidence will need to be taken into account, including preharvest factors [19,20]. Such a "nonpolluting integrated disease management" (NPIDM) concept should not be confused with the traditional "integrated disease management" (IDM) in the context of agricultural "integrated production," which often implies fruit production in compliance with particular national or regional regulations and programs that still include the use of postharvest fungicides. In Valencia (Spain), for instance, the regional administrative rule for IDM of citrus fruit allows the use of postharvest conventional fungicides "when necessary," and the only restriction is that they have to be applied "under technical supervision" [21].

The establishment of NPIDM strategies against target postharvest diseases is based on the knowledge of pathogen biology and epidemiology and needs to consider all preharvest, harvest, and postharvest factors that can influence disease appearance and incidence in order to minimize decay losses. Of course, all the actions planned, in the field or after harvest, should be cost-effective and not adversely affect fruit quality. It is clear in this context, especially in the case of diseases caused by wound pathogens, that the basis of successful NPIDM strategies is the commercial adoption of suitable nonpolluting postharvest antifungal treatments to replace the use of conventional postharvest fungicides. In general, according to their nature, these alternative treatments can be physical, chemical, or biological [6,19]. Chemical alternatives should be compounds with known and minimal toxicological effects on mammals and impact on the environment and, as substances that will be in contact with fresh produce, they should be affirmed as generally recognized as safe (GRAS) by the United States Food and Drug Administration (US FDA), as food additives by the European Food Safety Authority (EFSA), or as an equivalent status by national legislations of other countries. GRAS materials are exempt from residue tolerances on all agricultural commodities by the US FDA. Low-toxicity chemicals that are recognized as GRAS include some essential oils, plant extracts, and other natural compounds of varying composition, but also synthetic inorganic or organic salts such as carbonates, bicarbonates, sorbates, benzoates, acetates, paraben salts, silicates, etc. [22]. When compared to other GRAS compounds, the advantages of these salts are their great availability, ease of handling and use, and low cost. Research methods for evaluating the suitability of these kinds of salts as alternative treatments for the control of postharvest diseases of fresh fruits, as well as noteworthy research results and successful commercial applications, are described in this review.

2. Evaluation and Selection of GRAS Salts

Although the acidic forms of some GRAS salts can also show substantial antimicrobial activity, salt compounds are preferred as potential postharvest treatments because of their superior solubility and ease of manipulation and application. Moreover, the additional antifungal activity against important postharvest pathogens of cations such as Na+, K+, and NH4+ has been proven for many salts [23,24]. Research with GRAS salts to control fruit postharvest decay generally implies a sequential procedure. *In vitro* tests are useful to assess the toxicity of different salt concentrations to the target postharvest pathogen. Aqueous solutions of selected salts at selected concentrations can then be used in *in vivo* laboratory tests, with fruit artificially inoculated with pathogens to determine the control ability of the salt in conditions that resemble potential applications in the packinghouse. The commercial value of the potential implementation of postharvest treatments with selected salt solutions at selected concentrations can be tested afterwards in semicommercial or commercial trials. More recently, interest has grown with regard to the use of antifungal GRAS salts as ingredients of edible coatings. In the fresh fruit industry, such coatings are devoted for increased storage life, but also as a substitute for the prestorage waxes that are often applied in mixed formulations with conventional fungicides.

2.1 In Vitro Antifungal Activity

Given a postharvest disease of economic importance, its causal microorganism should be isolated from diseased fruit, purified, properly identified, and multiplied to be used in the experiments. Postharvest fungal pathogens are typically cultured and multiplied on Petri dishes containing potato dextrose agar (PDA) culture medium, although other media such as malt extract agar (MEA), V-8, or dichloran rose-bengal chloramphenicol agar (RBDC) may be used. Commonly, different strains of the same pathogen are isolated from infected fresh fruits found in commercial packinghouses in the commodity producing area, and preliminary *in vivo* tests are conducted to select, based on their aggressiveness and uniform behavior, those strains of each species more suitable for use in research. Appropriate incubation temperatures for growth of the most common postharvest pathogens of fresh fruit are between 20 and 25 °C. Depending on the growth rate of the fungal species, the incubation time needed to obtain mature fungal inoculum for use in experiments is 1–3 weeks.

The most typical procedure for *in vitro* evaluation of the antifungal activity of GRAS salts is to assess the inhibition of the radial mycelial growth of the target pathogen. For this, the fungus is inoculated on 90 mm diameter plastic Petri dishes with PDA medium amended with the test salt at the desired concentration. Since the solubility in water of most GRAS salts is very high, sterile stock solutions of each salt at high concentration (5–10%) are prepared and serial dilutions are performed to achieve the range of final concentrations that will be tested, frequently from 0.1% or lower to a maximum of 2–3%, depending on the salt characteristics and the regulations established for each of them. These solutions are then incorporated into the autoclaved PDA medium at 40–50 °C, poured into the Petri dishes, and allowed to solidify, all under strict sterile conditions. Spores at a known concentration or, more commonly, mycelial plugs (around 5 mm in diameter) from the edge of the pathogen growing cultures, produced with a sterilized cork borer, are then inoculated at the center of each dish. PDA plates without salt serve as negative controls. Inoculated plates are then incubated in a growth cabinet at 20–25 °C for a period of time that depends on the fungal species, but at least until fungal growth completely covers the control plates. Radial mycelial growth is periodically (every 1, 2, or 3 days) determined in each plate by measuring two perpendicular fungal colony diameters during the entire incubation period. Usually, 3–5 replicate plates are used for each salt and salt concentration. The results are expressed as a percentage of mycelial growth inhibition according to the formula $(dc - dt)/dc \times 100$, where dc is the average diameter of the fungal colony on control plates and dt is the average diameter of the fungal colony on treated (salt-amended) plates. The obtained data are typically subjected to a two-way analysis of variance (ANOVA) with salt and salt concentration as factors. With particular salts selected for their high antifungal activity, it can be useful to establish the minimum inhibitory concentration (MIC) of the salt, which implies testing a larger number of concentrations.

Other valuable information that can be obtained in *in vitro* tests is the ability of GRAS salts to kill or inactivate the spores of the target fungal pathogen. For this, typical *in vitro* spore mortality or spore germination tests consist of preparing liquid culture medium (PDA broth or similar) containing different concentrations of the GRAS salt, to which aliquots of a spore suspension of known density (usually 10^4–10^6 spores/mL) are aseptically transferred [24,25]. After 18–24 h of incubation at 20–25 °C, acid fuchsin solution is added to stop further germination and the percentage of germinated spores is determined by observing 100–150 spores with an inverted compound microscope with a micrometer. As a control, the same amount of spore suspension is added to medium broth without GRAS salt. A spore is scored as germinated if the germ tube length is equal to or exceeds that of the spore itself. The data are generally expressed as percent spore germination inhibition and calculated with a formula similar to that described above for the percent mycelial growth inhibition. Each treatment (each salt at a defined concentration) is applied to 3–5 replicates. Microwell plates are often used for this type of test. With particular salts selected for their high antifungal activity, it can be useful to determine the ED_{50} or ED_{95} values, i.e., the effective doses (salt concentrations) that kill or inactivate 50% or 95%, respectively, of the spores.

2.2 Control Ability of Aqueous Solutions

GRAS salts and concentrations that show significant antifungal activity in *in vitro* tests are selected for evaluation of their ability to control disease in *in vivo* tests, i.e., in the fruit host. The commercial potential of postharvest treatment with aqueous solutions of salts with antifungal properties is high, because their use in fresh produce packinghouses to substitute for conventional fungicides will not require substantial changes in the mode of application and in the industrial equipment used. Like fungicides, they could be applied in drenchers or in the packing line as dips, sprays, or low-pressure floods.

Different types of *in vivo* tests can be designed, depending on the objective and scale of each particular research step. Sequentially, it is common to start with *in vivo* primary screenings in the laboratory, continue with small-scale trials, and finish with semicommercial or commercial trials [26].

The scale and the amount of fruit employed in each of these types of experiments are larger than in the previous one. Fruits are usually collected from commercial orchards or from the packinghouse if no postharvest treatments have been applied yet. They are used in the experiments the same day or following days, or, depending on the commodity, they could be used after a variable, but not prolonged, period of cold storage. It is important before each experiment to properly select, randomize, wash and disinfect, and thoroughly rinse the fruit.

2.2.1. In Vivo Primary Screenings

This denomination is usually given to very small-scale laboratory tests designed to select the best GRAS salts and salt concentrations for a particular type of fruit species or cultivar. Previous information from *in vitro* tests is useful as a starting point, but experience shows that the efficacy of treatments applied to fruit often cannot be anticipated from the results obtained with pathogens growing in artificial media. Disease development in fruit is the result of complex interactions between host, pathogen, and environment. In the case of diseases caused by wound pathogens, the inhibitory ability of GRAS salts depends on the presence of residues within the wound infection sites occupied by the fungus and on interactions between this residue and constituents of the rind. Apparently, the nature of such interactions may alter the original toxicity of the salts to the pathogen, and therefore their control ability cannot be predicted by their activity *in vitro* [26,27]. For this reason, in some cases, the *in vitro* toxicity of substances is not preliminarily assessed and *in vivo* primary screenings are the first approach to select effective GRAS salts and concentrations.

In general, two kinds of antifungal activity can be assessed in in vivo primary screenings (and also in subsequent small-scale trials, explained below): curative activity, in which fruits artificially inoculated with pathogens are treated with the salt solution after different periods of time, and preventive or protective activity, in which fruits treated with salt solution are artificially inoculated with pathogens after different periods of time [28–30]. A period of 24 h is frequently used for diseases caused by wound pathogens, since it is similar to the time between the most frequent infections in the field, which occur at harvest time, and potential treatment in the packinghouse [31]. A highly concentrated sterile mother solution of the salt is prepared to obtain, by dilution, the desired concentrations to be tested. The salt solution is applied with a micropipette (30–50 μL) in wounds inflicted in the rinds of mature fruit a specific period of time after (for assessment of curative activity) or before (for assessment of preventive activity) artificial inoculation of the target pathogen in these rind wounds. Typically, conidial suspensions of known concentration (104 to 106 spores/mL) are prepared from young PDA fungal cultures (incubation of 7–21 days at 20–25°C) dispersed in Tween 80®, filtered through two layers of cheesecloth to separate hyphal fragments, and adjusted to the desired concentration using a hemocytometer. Fungal inoculation is performed by applying a small volume (10–30 μL) of the conidial suspension to the rind wound. In other cases, the rind wound and the inoculation are performed at the same time by immersing the tip of a sterile stainless steel rod in the conidial suspension and inserting it into the fruit rind. Control fruits are inoculated and treated with sterile distilled water. Depending on the fruit size and shape, one or more rind wounds per fruit can be inflicted. The usual sample size for this type of test is 2–4 replicates of 5–10 fruits each. Treated fruits are incubated at 20–25 °C and high relative humidity (RH) for a variable period of time (until most of control fruits are actually decayed), which can be approximately of 1–3 weeks at these incubation conditions. Disease incidence (percentage of infected wounds) and severity (lesion diameter, in mm or cm) and pathogen sporulation (percentage of lesions showing spores) are determined every 3–7 days. Results can be also expressed as percent reductions with respect to control fruit, especially if results from several independent experiments are compiled [31].

2.2.2. Small-Scale Trials

Small-scale trials are typically conducted with fresh fruit artificially inoculated with the target pathogen to establish the best postharvest treatment conditions that resemble potential commercial applications in fresh produce packinghouses. In general, dip treatments are evaluated for research purposes with aqueous solutions of GRAS salts, because, compared to other aqueous or wax applications, dipping is the most effective method. This is also true for the application of conventional postharvest fungicides [32]. However, in some cases, other application systems such as drenching, high-pressure spraying, or low-pressure rinsing can be tested. Depending on the experiment and the commodity, inoculated and treated fruit can be incubated at 20–25 °C to favor fast fungal development and obtain results in about 1–2 weeks. This procedure for evaluation of dips in solutions of GRAS salts for the control of postharvest diseases is illustrated in Figure 1. The example in the figure refers to citrus green mold caused by *P. digitatum*. In other experiments, fruit dipped in salt solution can be stored at low temperatures for longer periods to resemble commercial fruit handling. Again, both curative and preventive activities can be tested in this type of experiment. Since aqueous solutions of GRAS salts are often synergistic with heat for disease control [22,33], dip treatments of different lengths (30 s to 3 min) with solutions heated at different temperatures (45–60 °C) can be tested. For example, a stainless steel water tank fitted with electrical resistances and a thermostat, able to heat fruit placed in steel buckets, is used at the Centre de Tecnologia Postcollita of the Institut Valencià d'Investigacions Agràries (IVIA CTP) for this purpose (Figure 1) [34]. Control fruits are typically treated with water at 20 °C for 30–60 s. The usual sample size for these trials is 3–5 replicates of 20–25 fruits each. Treated fruits are commonly arranged in plastic cavity sockets on cardboard or plastic trays before incubation or cold storage. Disease incidence and severity and pathogen sporulation are periodically determined during these storage periods.

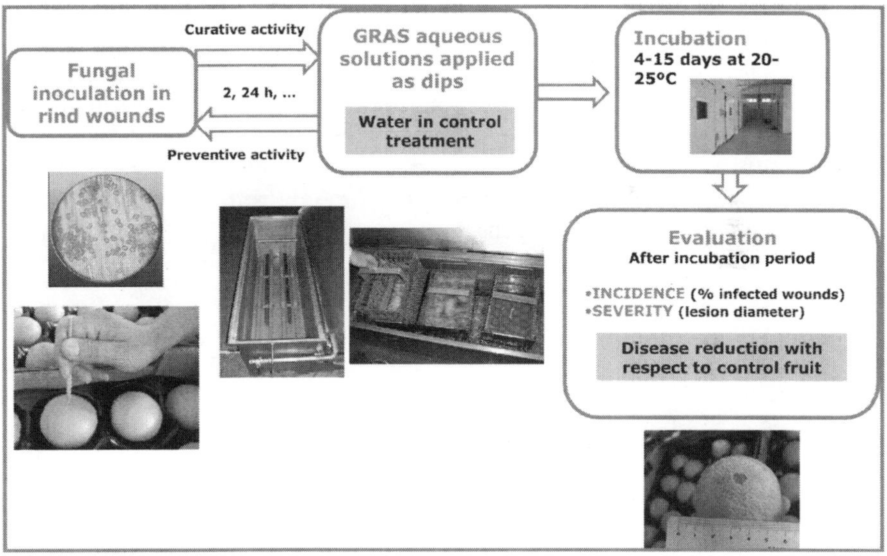

Figure 1. Methodological procedure for evaluating in small-scale trials the ability of dips in aqueous solutions of generally recognized as safe (GRAS) salts to control postharvest green mold of citrus caused by the fungus Penicillium digitatum.

2.2.3. Semicommercial or Commercial Trials

As a final conclusive step to assess the potential industrial value of GRAS salt application, semicommercial or commercial trials can be conducted with naturally infected fruits (not artificially inoculated with a target pathogen) and larger sample sizes than in previous trials. Pilot plants with research facilities and equipment, such as fruit packing lines or drenchers that are small-scale versions of those in commercial packinghouses, are used for semicommercial trials, with samples of varied

size depending on the type of fruit commodity, e.g., 3–5 replicates of 100–300 fruits per treatment. Commercial trials are directly conducted in industrial facilities with several replications per treatment of various entire fruit field boxes, bins, or packages. Besides the mandatory negative control treatment, which is fruit treated with water, it is frequent in these trials to add a positive control treatment consisting of the most common conventional postharvest fungicide, in order to compare the efficacy of the experimental treatments with that of current industrial practices. In every case, treated fruits are commercially handled in the packinghouse and follow the same postharvest handling steps as fruits used for actual commercialization (grading, cold storage, etc.). The performance of GRAS treatments is determined in terms of reduction of total decay, reduction of particular diseases, and effects on the quality of commercial produce.

2.2.4. Data Analysis

Dependent variables used to assess the effectiveness of GRAS salts in the above trials are disease incidence and severity and pathogen sporulation, or these variables expressed as percentages of reduction with respect to control fruit, which allow combining results from different analog trials. Depending on the experiment, additional variables that may be taken into account are incidence and severity of potential phytotoxicity induced to the fruit peel by the treatment, and fruit quality attributes during the cold storage period and after shelf life. Depending on the nature of the response variables, different statistical approaches can be selected for data analysis. A generalized linear model (GLM), such as binary logistic regression, or multifactor ANOVA followed by means separation tests are typically used.

2.3 Performance of Ingredients of Edible Coatings

Another application of antifungal GRAS salts that is increasingly gaining importance is as ingredients of edible coatings for fresh fruit. Postharvest application of these coatings could replace the use of fungicide-amended commercial waxes applied to many fruit commodities and could be used for organic or "zero-residue" fresh fruit production systems.

Waxes and coatings in general are primarily applied to fresh fruits to increase their postharvest life by regulating the exchange of water and gases (oxygen and carbon dioxide). This primary function allows for less weight loss during storage and, in some cases, the alleviation of some postharvest physiological disorders such as chilling injury or rind breakdown. In addition, coatings can also provide shine and gloss to improve the fruit's external appearance [35]. Besides natural coatings such as chitosan and *Aloe* spp. gels, most coatings are synthetic formulations comprising blends of hydrocolloids (proteins or polysaccharides) and lipids (waxes, acylglycerols, or fatty acids) as constituents of the composite coating matrix. These main ingredients are typically formulated with plasticizers (e.g., sucrose, glycerol, sorbitol, propylene glycol, etc.) and emulsifiers (e.g., fatty acids, polysorbates, monostearates, lecithin, etc.) to enhance the coating integrity and emulsion stability, respectively. Resins such as shellac are also often added to provide gloss [36,37]. Furthermore, additional ingredients can be incorporated to increase the functionality of these synthetic coatings. Among them, antimicrobial agents, such as some GRAS salts, can be added to prevent or reduce decay during storage [2,36].

The performance of antifungal edible coatings should be evaluated considering the interactions among all the elements of the system, i.e., the target pathogen, the coating ingredients, the antifungal GRAS salt, and the commodity and its postharvest handling. Therefore, developing these coatings requires initial optimization of coating formulations based on the chemical compatibility of the ingredients to achieve stable emulsions capable of forming homogeneous coatings. For a particular fruit commodity, the matrix and the non-antifungal ingredients of the coating are selected according to the coating's physiological activity in terms of reducing weight loss and increasing storability. For a particular pathosystem, GRAS salts and concentrations are initially selected in accordance with the control ability of aqueous solutions, hence it is important that previous information on these aspects is available. Coating formulation is usually optimized on the basis of percent total solid content, total lipid content, and GRAS salt concentration, but other parameters such as viscosity, pH, and wettability for the particular fruit commodity are also important [38]. Formulations that are stable after incorporation of the selected salt at the selected concentration in the selected coating will then be tested to determine their ability to control the target postharvest disease. Incompatible emulsions solidify or show phase separation or undesirable physical characteristics. It can happen that a particular salt is not compatible at all with a particular coating, or that it is compatible only at salt concentrations or coating total solid contents below a specific threshold. For example, among 470 emulsions formulated with a hydroxypropyl methylcellulose (HPMC)-lipid (beeswax and shellac) matrix and about 30 antifungal food additives (mostly GRAS salts) and mixtures at a large range of concentrations, only 25 emulsions were selected for their high stability. They contained 6 to 8% solid content, 50% (dry basis) total lipid content, and a maximum of 2.5% (wet basis) food preservative [38].

Although the antifungal activity of stable coatings with GRAS salts may be tested *in vitro* (disk diameter tests) by preparing dry disks of coating film and placing them on the surface of culture medium previously inoculated with spores of the target pathogen, it is more practical and usual to evaluate the control ability of the coatings in *in vivo* tests. Figure 2 represents a schematic diagram for this type of experiment, particularly for the evaluation of HPMC-lipid edible coatings containing GRAS salts for the control of black spot on cherry tomato. Fresh fruit samples are selected, washed, artificially inoculated with the target pathogen, and, after about 24 h (curative activity), coated with the different coating treatments and allowed to dry on a mesh screen. Coatings are usually applied by fruit immersion for brief periods (10–30 s) [39], but they can also be applied by pipetting a small amount of the emulsion (0.1–0.5 mL, depending on the commodity) onto each fruit and rubbing manually with gloved hands to mimic coating application in industrial packing line roller conveyors [40]. Control fruits are inoculated, but are uncoated or treated with coatings formulated without GRAS salts. Depending on the experiment and the commodity, inoculated and treated fruits can be incubated at 20–25 °C or cold-stored, similar to commercial postharvest handling. Sample size, dependent variables, and statistical analyses used for this kind of *in vivo* trial are equivalent to those described above for the *in vivo* evaluation of GRAS salt aqueous solutions.

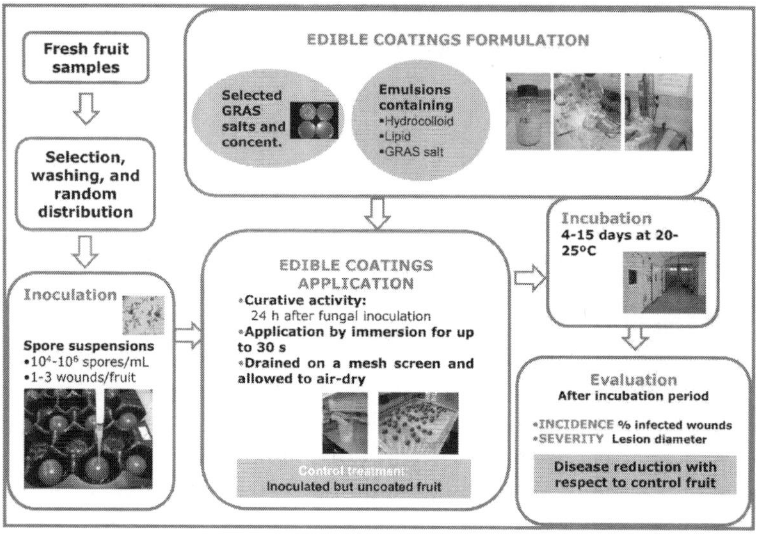

Figure 2. Methodological procedure for formulation and in vivo evaluation of the ability of edible coatings containing GRAS salts to control black spot of tomato caused by the fungus Alternaria alternata.

Once the coatings with the greatest ability to control disease are identified, and as a last step for the selection of the most feasible antifungal coatings for each particular application, it is very important to determine the effect of coating application on the physiological behavior and overall quality of coated fruit. For this purpose, both physicochemical and sensory fruit quality attributes are periodically evaluated during and after cold storage and simulated periods of shelf life at 20 °C [41]. Typically, the physicochemical fruit quality attributes that are assessed include weight loss (percent loss with respect to initial weight), fruit firmness (different types of measures with texturometers or penetrometers depending on the commodity), respiration and/or internal gas concentration (O_2 and CO_2 by gas chromatography), and overmaturation volatiles (ethanol and acetaldehyde contents by gas chromatography). Sensory fruit attributes such as flavor, off-flavors, and external and internal visual appearance should be evaluated by several trained judges with expertise in each particular commodity. In some cases, consumer tests by some nontrained individuals can also be of value.

3. Noteworthy Research and Commercial Results Obtained with GRAS Salts

Our research group at the IVIA CTP has worked for many years on evaluating GRAS salts for the control of fresh fruit postharvest decay. *In vitro* tests for preliminary selection of GRAS salts with activity against strains of *B. cinerea* and *A. alternata* pathogenic to cherry tomato fruit were performed in collaboration with researchers from Brazil [42], and against strains of *M. fructicola* pathogenic to plum fruit in collaboration with researchers from Turkey (Table 1) [43]. Most of the group's work, however, has focused on evaluating the effectiveness of aqueous solutions to control postharvest diseases of fresh fruit of economic importance in the Mediterranean area of the Iberian Peninsula. More recently, we started a research line devoted to developing and selecting edible coatings containing antifungal GRAS salts effective for the control of major postharvest diseases of citrus, stone fruits, tomato, persimmon, and pomegranate.

Table 1. Percentage inhibition of radial growth of Monilinia fructicola on potato dextrose agar (PDA) Petri dishes amended with different concentrations of GRAS salts after 7 days of incubation at 25°C.

	Inhibition of *Monilinia fructicola* (%) 1		
GRAS Salt	**Salt Concentration (%, *w/v*)**		
	0.2	**1.0**	**2.0**
Ammonium bicarbonate	100.00 iA	100.00 eA	100.00 cA
Potassium carbonate	81.76 gA	100.00 eB	100.00 cB
Potassium bicarbonate	89.22 hA	98.01 dB	100.00 cC
Potassium silicate	11.08 bA	100.00 eB	100.00 cB
Potassium sorbate	49.42 efA	100.00 eB	100.00 cB
Sodium carbonate	96.37 hiA	100.00 eB	100.00 cB
Sodium bicarbonate	100.00 iA	100.00 eA	100.00 cA
Sodium acetate	22.64 cA	62.21 bB	93.98 bC
Sodium diacetate	35.83 dA	100.00 eB	100.00 cB
Sodium benzoate	42.34 deA	91.92 cB	99.67 cC
Sodium formate	0.00 aA	60.44 aB	92.50 aC
Sodium propionate	54.37 fA	100.00 eB	100.00 cB
Sodium methylparaben 2	24.85 cA	100.00 eB	100.00 cB
Sodium ethylparaben 2	27.80 cA	100.00 eB	100.00 cB

Note: Means in rows with different capital letters and means in columns with different lowercase letters are significantly different by Fisher's protected least significant difference (LSD) test (P < 0.05) applied after ANOVA. 1 Colony diameter reduction with respect to control treatments (nonamended PDA plates). 2 Doses of the agents tested were 0.01, 0.05, and 0.1%. Reproduced from Karaca et al. (2014) [43] with permission from Elsevier.

3.1 Aqueous Solutions

Research on the evaluation of aqueous solutions of GRAS inorganic and organic salts for the control of postharvest diseases of fresh horticultural produce has been thoroughly reviewed recently, focusing especially on citrus fruit, temperate fruit, tropical fruit, and vegetables [22]. Since then, more interesting works on the subject have been published. The GRAS salt calcium chloride significantly reduced Rhizopus rot of peaches, and the reduction was greater when combined with lemongrass oil [44]. Although not a salt, vapors of the GRAS compound acetic acid were effective for reducing gray mold of stored table grapes caused by *B. cinerea* [45]. Among a wide range of organic acids and salts tested *in vitro* and *in vivo* for the control of gray mold of tomato, the salts potassium carbonate, sodium bicarbonate, sodium carbonate, sodium metabisulfite, and sodium salicylate were selected for their higher control ability on fruit artificially inoculated with *B. cinerea* [25]. Sodium bicarbonate also satisfactorily controlled black rot of yellow pitahaya caused by *A. alternata* [46]. Further studies on the mode of action of this salt to control citrus green mold caused by *P. digitatum* were recently released [47]. Heated solutions of sodium benzoate were effective for the reduction of green mold on artificially inoculated oranges either incubated at 20 °C or cold-stored at 5 °C [48].

Arguably, one field where the use of GRAS salt solutions has been commercially implemented with greater success is the control of citrus postharvest green and blue molds caused by *P. digitatum* and *P. italicum*, respectively. Taking into account early initial data from the 1920s in California [49], intense research was conducted in the 1990s and 2000s to set the basis of commercial dips with sodium carbonate and sodium bicarbonate to reduce or replace the use of synthetic fungicides in citrus packinghouses. It was found that 2–3% (*w*/*v*) dip treatments for 60–150 s were effective, although with differences, on oranges, lemons, and mandarins. These treatments were synergistic with heat, and solutions heated to 40–55 °C were more effective yet nonphytotoxic [24,33,50,51]. For instance, the influence of solution temperature, salt concentration, and immersion period on disease incidence is shown in Figure 3 for control with sodium carbonate of blue mold caused by *P. italicum* on Valencia oranges. Since then, and after several successful commercial applications attempted in California, sodium carbonate and sodium bicarbonate have been the most

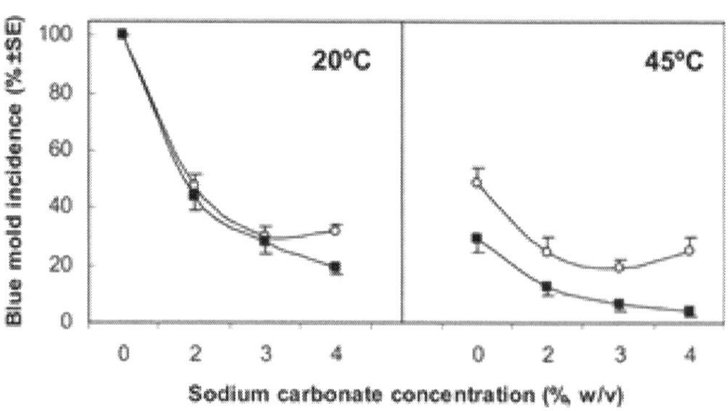

Figure 3. Influence of solution temperature, sodium carbonate concentration, and immersion period (O = 60 s, ■ = 150 s) on the incidence of blue mold on cv. Valencia oranges artificially inoculated with Penicillium italicum 24 h before treatment, rinsed at low pressure, and stored at 20°C and 90% relative humidity (RH) for 7 days. Data are the means of two experiments with five replicates of 25 fruit each. Reproduced from Palou et al. (2001) [51] with permission from the American Phytopathological Society.

common GRAS salts used for decay control in citrus packinghouses worldwide. Currently, the most usual mode of application is by dipping the fruit in large tanks containing heated solutions of the salts. However, other applications, such as the use of salt solutions in hot water rinsing and brushing (HWRB) equipment [52,53], have also been successfully attempted. In particular, satisfactory decay control was achieved on organic citrus fruit destined for export markets in Europe that had been treated in HWRB machinery in packinghouses in Israel with 2% sodium bicarbonate solution at 52 °C for 15 s (Fallik, personal communication). The production of electrolyzed sodium bicarbonate is another technology, assayed in Italy, that proved to be superior to both electrolyzed water and nonelectrolyzed sodium bicarbonate (regular salt aqueous solution) for the control of citrus green mold [54].

Besides sodium carbonates, another GRAS salt that has repeatedly shown good ability to control citrus postharvest green and blue molds is potassium sorbate. This organic salt has been successfully applied alone or in combination with commercial fungicides in relatively long dip applications (2–3 min) at room or warm temperature [26,55] or in shorter dip applications at higher temperatures [34,56]. As in the case of carbonates, potassium sorbate solutions were more effective when applied at high temperatures. Moreover, among a large variety of GRAS salts that were evaluated against the most important postharvest fungal diseases of stone fruit, potassium sorbate was the most effective, particularly for the control of brown rot caused by *M. fructicola* [57]. Compared to carbonate solutions, potassium sorbate solutions have the advantage of fewer disposal issues due to the absence of sodium and lower salinity and pH, but the disadvantages of higher price and greater phytotoxicity risk (fruit rind staining) if not well applied [34,56]. Probably for these reasons, the commercial use of potassium sorbate aqueous solutions in citrus packinghouses is much less than that of carbonate solutions. However, as discussed in the next subsection, this salt is more used as an antifungal ingredient of fruit coatings.

3.2 Edible Coatings

Substantial effort has been devoted in recent years to the development and selection of novel synthetic composite coatings containing GRAS salts as antimicrobial ingredients. Results with such coatings applied for alternative control of postharvest diseases have recently been summarized for fresh produce in general [22] and citrus fruit in particular [2]. Work by our group at the IVIA CTP showed that selected stable HPMC-lipid (beeswax and shellac) coatings formulated with salts or mixtures at concentrations of 0.05–4.5% (wet basis) substantially reduced green and blue molds on artificially inoculated orange and mandarin cultivars of commercial importance coated and either incubated at 20 °C for 7 days [39,58] (Figure 4) or cold-stored at 5 °C for 30–60 days plus 7 days of shelf life simulation at 20 °C [41,59]. Although the mold control ability of the coatings depended on the citrus species and cultivar, in general those containing the salts potassium sorbate, sodium benzoate, sodium propionate, or mixtures of them were the most effective. In further studies, the physiological performance of these selected coatings was tested during long-term cold storage of noninoculated fruit, and overall, the coatings significantly reduced weight loss and maintained firmness of coated oranges and mandarins.

Although the coatings modified the gas composition of the internal atmosphere of coated fruit, they did not affect the fruit sensory quality [41,59]. Furthermore, some of these coatings are also being evaluated with satisfactory results for the control of other important postharvest diseases of citrus such as anthracnose and stem-end rot caused by the fungi *C. gloeosporioides* and *L. theobromae*, respectively (Palou, unpublished).

Other similar studies conducted by our group demonstrated the ability of different HPMC-lipid–based coatings containing GRAS salts to control gray mold and black spot of cherry tomato caused, respectively, by *B. cinerea* and *A. alternata* [42]. Coatings containing sodium propylparaben, sodium benzoate, or ammonium carbonate salt were the most effective and also reduced weight loss and maintained firmness of cherry tomatoes during cold storage [60,61]. Similarly, among a wide variety of antifungal GRAS salts tested as ingredients of HPMC-beeswax edible coatings, paraben salts and potassium sorbate were found to be the best for effective control of brown rot caused by *M. fructicola* on artificially inoculated plums [43]. These coatings also effectively reduced weight loss and maintained firmness and overall quality of coated plums during prolonged cold storage [40]. In general, potassium sorbate is the GRAS salt that has shown the best results for fruit decay control when incorporated into matrices of novel edible coatings or when mixed with conventional postharvest waxes [62]. For this reason, several commercial formulations containing this food additive have been registered and are currently available on the market for postharvest treatment of fresh fruit, especially citrus and stone fruits. Citrashine Plus® and Citrashine Plus Nature® from the manufacturer Decco Iberica Postcosecha S.A.U. (Paterna, Valencia, Spain) and Greengard-SK 50® from the manufacturer Fomesa Fruitech S.L. (Beniparrell, Valencia, Spain) are examples of this type of product commercialized in Spain.

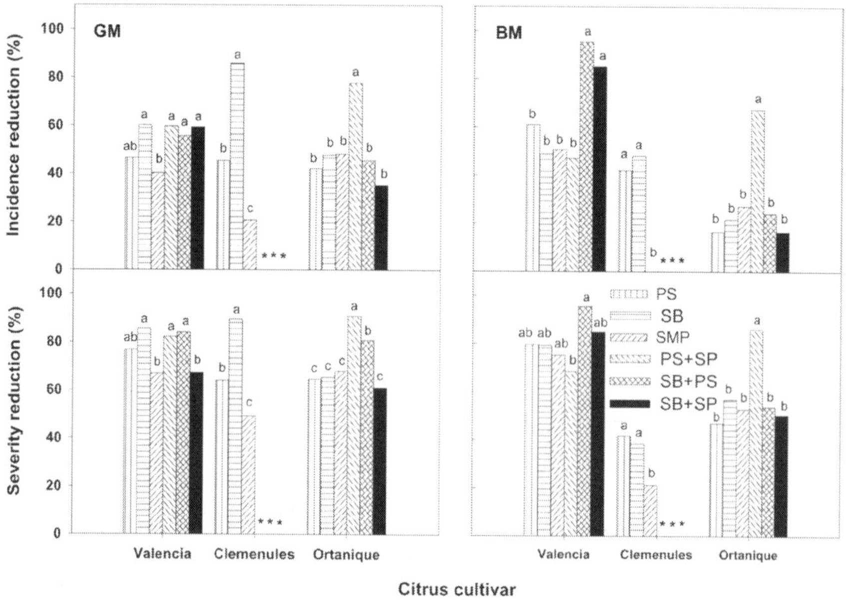

Figure 4. Curative activity of hydroxypropyl methylcellulose (HPMC)-lipid edible composite coatings, containing potassium sorbate (PS), sodium benzoate (SB),

sodium propionate (SP), sodium methylparaben (SMP), or a mixture of these, against green (GM) and blue (BM) molds on different citrus cultivars artificially inoculated with Penicillium digitatum and P. italicum, coated 24 h later, and incubated for 7 days at 20°C. Disease incidence and severity reductions were determined with respect to control fruit (inoculated but uncoated). For each mold and cultivar, columns with different letters are different by Fisher's protected LSD test (P < 0.05). * No data available. Reproduced from Palou et al. (2014) [58] with permission from the International Society for Horticultural Science.

References

1. Smilanick, J.L.; Brown, G.E.; Eckert, J.W. The biology and control of postharvest diseases. In *Fresh Citrus Fruits*, 2nd ed.; Wardowski, W.F., Miller, W.M., Hall, D.J., Grierson, W., Eds.; Florida Science Source: Longboat Key, FL, USA, 2006; pp. 339–396. ISBN 0-944961-08-8.

2. Palou, L.; Valencia-Chamorro, S.A.; Pérez-Gago, M.B. Antifungal edible coatings for fresh citrus fruit: A review. *Coatings* **2015**, *5*, 962–986. [CrossRef]

3. Louw, J.P.; Korsten, L. Pathogenic *Penicillium* spp. on apple and pear. *Plant Dis.* **2014**, *98*, 590–598. [CrossRef]

4. Errampalli, D. *Penicillium expansum* (Blue mold). In *Postharvest Decay. Control Strategies*; Bautista-Baños, S., Ed.; Academic Press, Elsevier Inc.: London, UK, 2014; pp. 189–231. ISBN 978-0-12-411552-1.

5. Zhang, M.; Xu, L.; Zhang, L.; Guo, Y.; Qi, X.; He, L. Effects of quercetin on postharvest blue mold control in kiwifruit. *Sci. Hortic.* **2018**, *228*, 18–25. [CrossRef]

6. Palou, L. *Penicillium digitatum, Penicillium italicum* (Green mold, Blue mold). In *Postharvest Decay. Control Strategies*; Bautista-Baños, S., Ed.; Academic Press, Elsevier Inc.: London, UK, 2014; pp. 45–102. ISBN 978-0-12-411552-1.

7. Barkai-Golan, R. Postharvest Diseases of Fruit and Vegetables. In *Development and Control*; Elsevier Science: Amsterdam, The Netherlands, 2001; ISBN 978-0-44-450584-2.

8. Bautista-Baños, S.; Bosquez-Molina, E.; Barrera-Necha, L.L. *Rhizopus stolonifer* (Soft rot). In *Postharvest Decay. Control Strategies*; Bautista-Baños, S., Ed.; Academic Press, Elsevier Inc.: London, UK, 2014; pp. 1–44. ISBN 978-0-12-411552-1.

9. Saito, S.; Xiao, C.L. Prevalence of postharvest diseases of mandarin fruit in California. *Plant Health Progress* **2017**, *18*, 204–210. [CrossRef]

10. Droby, S.; Lichter, A. Post-harvest *Botrytis* infection: Etiology, development and management. In *Botrytis: Biology, Pathology and Control*; Elad, Y., Williamson, B., Tudzynski, P., Delen, N., Eds.; Kluwer Academic Publishers: Dordrecht, The Netherlands, 2004; pp. 349–367. ISBN 1-4020-2624-2.

11. Romanazzi, G.; Feliziani, E. *Botrytis cinerea* (Gray mold). In *Postharvest Decay. Control Strategies*; Bautista-Baños, S., Ed.; Academic Press, Elsevier Inc.: London, UK, 2014; pp. 131–146. ISBN 978-0-12-411552-1.

12. Prusky, D.; Freeman, S.; Dickman, M.B. (Eds.) Colletotrichum. In *Host Specificity, Pathology, and Host -Pathogen Interaction*; APS Press: St. Paul, MN, USA, 2000; ISBN 0-89054-258-9.
13. Siddiqui, Y.; Ali, A. *Colletotrichum gloeosporioides* (Anthracnose). In *Postharvest Decay. Control Strategies*; Bautista-Baños, S., Ed.; Academic Press, Elsevier Inc.: London, UK, 2014; pp. 337–371. ISBN 978-0-12-411552-1.
14. Rotem, J. The Genus Alternaria. In *Biology, Epidemiology and Pathogenicity*; APS Press: St. Paul, MN, USA, 1994; ISBN 0-89054-152-3.
15. Zhang, J. *Lasiodiplodia theobromae* in citrus fruit (Diplodia stem-end rot). In *Postharvest Decay. Control Strategies*; Bautista-Baños, S., Ed.; Academic Press, Elsevier Inc.: London, UK, 2014; pp. 309–335. ISBN 978-0-12-411552-1.
16. Troncoso-Rojas, R.; Tiznado-Hernández, M.E. *Alternaria alternata* (Black rot, black spot). In *Postharvest Decay. Control Strategies*; Bautista-Baños, S., Ed.; Academic Press, Elsevier Inc.: London, UK, 2014; pp. 147–187. ISBN 978-0-12-411552-1.
17. Martini, C.; Mari, M. *Monilinia fructicola, Monilinia laxa* (Monilinia rot, brown rot). In *Postharvest Decay. Control Strategies*; Bautista-Baños, S., Ed.; Academic Press, Elsevier Inc.: London, UK, 2014; pp. 233–265. ISBN 978-0-12-411552-1.
18. Zhu, X.; Niu, C.; Chen, X.; Guo, L. *Monilinia* species associated with brown rot of cultivated apple and pear fruit in China. *Plant Dis.* **2016**, *100*, 2240–2250. [CrossRef]
19. Palou, L.; Smilanick, J.L.; Droby, S. Alternatives to conventional fungicides for the control of citrus postharvest green and blue moulds. *Stewart Postharv. Rev.* **2008**, *4*, 1–16. [CrossRef]
20. Wisniewski, M.; Droby, S.; Norelli, J.; Liu, J.; Schena, L. Alternative management technologies for postharvest disease control: The journey from simplicity to complexity. *Postharvest Biol. Technol.* **2016**, *122*, 3–10. [CrossRef]
21. Palou, L. Control integrado no contaminante de enfermedades de poscosecha (CINCEP): Nuevo paradigma para el sector español de los cítricos. *Levante Agrícola* **2011**, *406*, 173–183.
22. Palou, L.; Ali, A.; Fallik, E.; Romanazzi, G. GRAS, plant- and animal-derived compounds as alternatives to conventional fungicides for the control of postharvest diseases of fresh horticultural produce. *Postharvest Biol. Technol.* **2016**, *122*, 41–52. [CrossRef]
23. Palmer, C.L.; Horst, R.K.; Langhans, R.W. Use of bicarbonates to inhibit in vitro colony growth of *Botrytis cinerea*. *Plant Dis.* **1997**, *81*, 1432–1438. [CrossRef]
24. Smilanick, J.L.; Margosan, D.A.; Mlikota-Gabler, F.; Usall, J.; Michael, I.F. Control of citrus green mold by carbonate and bicarbonate salts and the influence of commercial postharvest practices on their efficacy. *Plant Dis.* **1999**, *83*, 139–145. [CrossRef]
25. Alaoui, F.T.; Askarne, L.; Boubaker, H.; Boudyach, E.H.; Aoumar, A.A.B. Control of gray mold disease of tomato by postharvest application of organic acids and salts. *Plant Pathol. J.* **2017**, *16*, 62–72. [CrossRef]

26. Palou, L.; Usall, J.; Smilanick, J.L.; Aguilar, M.J.; Viñas, I. Evaluation of food additives and low-toxicity compounds as alternative chemicals for the control of *Penicillium digitatum* and *Penicillium italicum* on citrus fruit. *Pest Manag. Sci.* **2002**, *58*, 459–466. [CrossRef] [PubMed]
27. Hwang, L.; Klotz, L.J. The toxic effect of certain chemical solutions on spores of *Penicillium italicum* and *P. digitatum*. *Hilgardia* **1938**, *12*, 1–38. [CrossRef]
28. Palou, L.; Marcilla, A.; Rojas-Argudo, C.; Alonso, M.; Jacas, J.A.; del Río, M.A. Effects of X-ray irradiation and sodium carbonate treatments on postharvest *Penicillium* decay and quality attributes of clementine mandarins. *Postharvest Biol. Technol.* **2007**, *46*, 252–261. [CrossRef]
29. Moscoso-Ramírez, P.A.; Montesinos-Herrero, C.; Palou, L. Characterization of postharvest treatments with sodium methylparaben to control citrus green and blue molds. *Postharvest Biol. Technol.* **2013**, *77*, 128–137. [CrossRef]
30. Moscoso-Ramírez, P.A.; Palou, L. Preventive and curative activity of postharvest potassium silicate treatments to control green and blue molds on orange fruit. *Eur. J. Plant Pathol.* **2014**, *138*, 721–732. [CrossRef]
31. Eckert, J.W.; Brown, G.E. Evaluation of postharvest fungicide treatments for citrus fruits. In *Methods for Evaluating Pesticides for Control of Plant Pathogens*; Hickey, K.D., Ed.; APS Press: St. Paul, MN, USA, 1986; pp. 92–97. ISBN 0-89054-071-3.
32. Smilanick, J.L.; Michael, I.F.; Mansour, M.F.; Mackey, B.E.; Margosan, D.A.; Flores, D.; Weist, C.F. Improved control of green mold of citrus with imazalil in warm water compared with its use in wax. *Plant Dis.* **1997**, *81*, 1299–1304. [CrossRef]
33. Palou, L.; Smilanick, J.L.; Usall, J.; Viñas, I. Control of postharvest blue and green molds of oranges by hot water, sodium carbonate, and sodium bicarbonate. *Plant Dis.* **2001**, *85*, 371–376. [CrossRef]
34. Montesinos-Herrero, C.; del Río, M.A.; Pastor, C.; Brunetti, O.; Palou, L. Evaluation of brief potassium sorbate dips to control postharvest penicillium decay on major citrus species and cultivars. *Postharvest Biol. Technol.* **2009**, *52*, 117–125. [CrossRef]
35. Pérez-Gago, M.B.; Palou, L. Antimicrobial packaging for fresh and fresh-cut fruits and vegetables. In *Fresh-Cut Fruits and Vegetables: Technology, Physiology, and Safety*; Pareek, S., Ed.; CRC Press: Boca Raton, FL, USA, 2016; pp. 403–452. ISBN 978-1-49-872994-9.
36. Valencia-Chamorro, S.A.; Pérez-Gago, M.B.; del Río, M.A.; Palou, L. Antimicrobial edible films and coatings for fresh and minimally processed fruits and vegetables: A review. *Crit. Rev. Food Sci. Nutr.* **2011**, *51*, 872–900. [CrossRef] [PubMed]
37. Han, J.H. Edible films and coatings: A review. In *Innovations in Food Packaging*; Han, J.H., Ed.; Academic Press, Elsevier Inc.: Amsterdam, The Netherlands, 2014; pp. 213–255. ISBN 978-0-12-394601-0.
38. Valencia-Chamorro, S.A.; Palou, L.; del Rio, M.A.; Pérez-Gago, M.B. Inhibition of *Penicillium digitatum* and *Penicillium italicum* by hydroxypropyl methylcellulose-lipid edible composite films containing food additives with

antifungal properties. *J. Agric. Food Chem.* **2008**, *56*, 11270–11278. [CrossRef] [PubMed]

39. Valencia-Chamorro, S.A.; Pérez-Gago, M.B.; del Río, M.A.; Palou, L. Curative and preventive activity of hydroxypropyl methylcellulose-lipid edible composite coating antifungal food additives to control citrus postharvest green and blue molds. *J. Agric. Food Chem.* **2009**, *57*, 2770–2777. [CrossRef] [PubMed]

40. Gunaydin, S.; Karaca, H.; Palou, L.; de la Fuente, B.; Pérez-Gago, M.B. Effect of hydroxypropyl methylcellulose-beeswax composite edible coatings formulated with or without antifungal agents on physicochemical properties of plums during cold storage. *J. Food Qual.* **2017**, 8573549. [CrossRef]

41. Valencia-Chamorro, S.A.; Pérez-Gago, M.B.; del Río, M.A.; Palou, L. Effect of antifungal hydroxypropyl methylcellulose (HPMC)-lipid edible composite coatings on penicillium decay development and postharvest quality of cold-stored 'Ortanique' mandarins. *J. Food Sci.* **2010**, *75*, 418–426. [CrossRef] [PubMed]

42. Fagundes, C.; Pérez-Gago, M.B.; Monteiro, A.R.; Palou, L. Antifungal activity of food additives in vitro and as ingredients of hydroxypropyl methylcellulose-lipid edible coatings against *Botrytis cinerea* and *Alternaria alternata* on cherry tomato fruit. *Int. J. Food Microbiol.* **2013**, *166*, 391–398. [CrossRef] [PubMed]

43. Karaca, H.; Pérez-Gago, M.B.; Taberner, V.; Palou, L. Evaluating food additives as antifungal agents against *Monilinia fructicola* in vitro and in hydroxypropyl methylcellulose-lipid composite edible coatings for plums. *Int. J. Food Microbiol.* **2014**, *179*, 72–79. [CrossRef] [PubMed]

44. Salem, E.A.; Youssef, K.; Sanzani, S.M. Evaluation of alternative means to control postharvest *Rhizopus* rot of peaches. *Sci. Hortic.* **2016**, *198*, 86–90. [CrossRef]

45. Venditti, T.; Ladu, G.; Cubaiu, L.; Myronycheva, O.; D'hallewin, G. Repeated treatments with acetic acid vapors during storage preserve table grapes fruit quality. *Postharvest Biol. Technol.* **2017**, *125*, 91–98. [CrossRef]

46. Vilaplana, R.; Alba, P.; Valencia-Chamorro, S. Sodium bicarbonate salts for the control of postharvest black rot disease in yellow pitahaya (*Selenicereus megalanthus*). *Crop Prot.* **2018**, *114*, 90–96. [CrossRef]

47. Venditti, T.; D'hallewin, G.; Ladu, G.; Petretto, G.L.; Pintore, G.; Labavitch, J.M. Effect of $NaHCO_3$ treatments on the activity of cell-wall-degrading enzymes produced by *Penicillium digitatum* during the pathogenesis process on grapefruit. *J. Sci. Food Agric.* **2018**, *98*, 4928–4936. [CrossRef] [PubMed]

48. Palou, L.; Moscoso-Ramírez, P.A.; Montesinos-Herrero, C. Assessment of optimal postharvest treatment conditions to control green mold of oranges with sodium benzoate. *Acta Hortic.* **2018**, *1194*, 221–226. [CrossRef]

49. Barger, W.R. Sodium bi-carbonate as citrus fruit disinfectant. *Calif. Citrogr.* **1928**, *13*, 172–174.

50. Smilanick, J.L.; Mackey, B.E.; Reese, R.; Usall, J.; Margosan, D.A. Influence of concentration of soda ash, temperature, and immersion period on the control of postharvest green mold on oranges. *Plant Dis.* **1997**, *81*, 379–382. [CrossRef]

51. Palou, L.; Usall, J.; Muñoz, J.A.; Smilanick, J.L.; Viñas, I. Hot water, sodium carbonate, and sodium bicarbonate for the control of postharvest green and blue molds of clementine mandarins. *Postharvest Biol. Technol.* **2002**, *24*, 93–96. [CrossRef]

52. Fallik, E. Prestorage hot water treatments (immersion, rinsing and brushing). *Postharvest Biol. Technol.* **2004**, *32*, 125–134. [CrossRef]

53. Fallik, E. Hot water treatments of fruits and vegetables for postharvest storage. *Hortic. Revs.* **2010**, *38*, 191–212. [CrossRef]

54. Fallanaj, F.; Ippolito, A.; Ligorio, A.; Garganese, F.; Zavanella, C.; Sanzani, S.M. Electrolyzed sodium bicarbonate inhibits *Penicillium digitatum* and induces defence responses against green mould in citrus fruit. *Postharvest Biol. Technol.* **2016**, *115*, 18–29. [CrossRef]

55. Hall, D.J. Comparative activity of selected food preservatives as citrus postharvest fungicides. *Proc. Fla. State Hort. Soc.* **1988**, *101*, 184–187.

56. Smilanick, J.L.; Mansour, M.F.; Mlikota-Gabler, F.; Sorenson, D. Control of citrus postharvest green mold and sour rot by potassium sorbate combined with heat and fungicides. *Postharvest Biol. Technol.* **2008**, *47*, 226–238. [CrossRef]

57. Palou, L.; Smilanick, J.L.; Crisosto, C.H. Evaluation of food additives as alternative or complementary chemicals to conventional fungicides for the control of major postharvest diseases of stone fruit. *J. Food Prot.* **2009**, *72*, 1037–1046. [CrossRef] [PubMed]

58. Palou, L.; Valencia-Chamorro, S.A.; Pérez-Gago, M.B. Edible composite coatings formulated with antifungal GRAS compounds: A novel approach for postharvest preservation of fresh citrus fruit. *Acta Hortic.* **2014**, *1053*, 143–149. [CrossRef]

59. Valencia-Chamorro, S.A.; Palou, L.; Del Río, M.A.; Pérez-Gago, M.B. Performance of hydroxypropyl methylcellulose (HPMC)-lipid edible composite coatings containing food additives with antifungal properties during cold storage of 'Clemenules' mandarins. *LWT Food Sci. Technol.* **2011**, *44*, 2342–2348. [CrossRef]

60. Fagundes, C.; Palou, L.; Monteiro, A.R.; Pérez-Gago, M.B. Effect of antifungal hydroxypropyl methylcellulose-beeswax edible coatings on gray mold development and quality attributes of cold-stored cherry tomato fruit. *Postharvest Biol. Technol.* **2014**, *92*, 1–8. [CrossRef]

61. Fagundes, C.; Palou, L.; Monteiro, A.R.; Pérez-Gago, M.B. Hydroxypropyl methylcellulose-beeswax edible coatings formulated with antifungal food additives to reduce alternaria black spot and maintain postharvest quality of cold-stored cherry tomatoes. *Sci. Hortic.* **2015**, *193*, 249–257. [CrossRef]

62. Youssef, K.; Ligorio, A.; Nigro, F.; Ippolito, A. Activity of salts incorporated in wax in controlling postharvest diseases of citrus fruit. *Postharvest Biol. Technol.* **2012**, *65*, 39–43. [CrossRef]

Index